THE LOGIC OF AFFECT

ALSO BY PAUL REDDING

Hegel's Hermeneutics

The Logic
of Affect

PAUL REDDING

CORNELL UNIVERSITY PRESS

ITHACA AND LONDON

First published 1999 by Cornell University Press

Library of Congress Cataloging-in-Publication Data
Redding, Paul, b. 1948
    The logic of affect / Paul Redding.
      p. cm.
    Includes bibliographical references and index.
    ISBN 0-8014-3591-9 (cloth : alk. paper)
    1. Emotions (Philosophy)—History.   2. Emotions and cognition—
History. I. Title.
    B815.R43   1999
    128'.37—dc21   98-55361

Printed in the United States of America

Cloth printing   10 9 8 7 6 5 4 3 2 1

FSC FSC Trademark © 1996 Forest Stewardship Council A.C.
SW-COC-098

Contents

3
FREUD, AFFECT, AND THE LOGIC OF THE UNCONSCIOUS
46

4
KANT, MIND, AND SELF-CONSCIOUSNESS
71

5
THE UNSAYABLE SELF-FEELING BODY: FEELING, REPRESENTATION, AND REALITY IN FICHTE'S TRANSCENDENTAL IDEALISM
88

Acknowledgments

I thank the School of Philosophy at the University of Sydney for providing me with valuable leave from teaching duties during 1996–1997 and Professor Pierre Beumont in the Department of Psychological Medicine for generously providing me with a welcoming environment during this time, allowing this work to get under way. My thanks also goes to the School of Advanced Study at the University of London and its director, Professor Jonathan Wolff, for accepting me as a research fellow into that stimulating environment in late 1996.

A great number of people have been very helpful in steering me around all sorts of issues germane to the logic of affect and for providing valuable feedback. I can in no way thank all of them individually, but let me mention Damian Byers, Keith Campbell, Jack Copeland, Paul Crittenden, Bill Fulford, Sebastian Gardner, Moira Gatens, Phillip Gerrans, Grant Gillett, Jim Hopkins, Stephen Houlgate, Eric Tsui James, Simon Lumsden, György Markus, Adrian McKenzie, Russell Meares, Peter Murray, Huw Price, Alison Ross, John Sutton, Stephanie Winfield, and members of the Philosophy Study Group of the NSW Institute of Psychiatry.

Once again I am especially grateful to Roger Haydon at Cornell University Press for his encouragement, guidance, and good spirits during the passage of this work from an idea and title to a completed manuscript, as I am to the press's three anonymous readers whose critical feedback was invaluable. Thanks also to Marie-Josée Schorp for her fine copy-editing.

Without the love, support, counsel, and endorphin-inducing presence of my soul mate, mate, and grooming partner, Vicki Varvaressos, neither this work, nor little else, would have been done.

This book is dedicated to the memory of my father, Robert Charles Redding, an affectionate man.

P. R.

Sydney, N.S.W.

THE LOGIC OF AFFECT

Introduction:
A Logic for the Reasons of the Heart?

C reating an aphorism that would prove irresistible to many later investigators into affective life, Pascal wrote: "The heart has its reasons,
which reason does not know."[1] Words such as these have recommended
themselves to those wishing to challenge the polarization of reason and
feeling typical of the strongly intellectualist culture of the West. For them
these words mean that the heart, *too*, has reason, and this fact is not recognized by reason itself. In particular, in the second half of the twentieth century words such as these would typically be invoked by those advocating
what has come to be known as a "cognitivist" view of the emotions. The
emotions, according to such an approach, should not be thought of as mere
"dumb" feelings. They are shot through with "cognitive" elements—
ideas, concepts, beliefs, interpretations, appraisals. But Pascal's words do
not say the heart has reason *too*. Rather, he says the heart has *its* reasons,
reasons not recognized by reason—suggesting some kind of difference between the reasons of the heart and the reasons of the head.

The guiding questions of this book orient themselves around this latter
gloss on Pascal's statement. If the heart does have its reasons, what are the
principles of their organization? And if these reasons are not the same as
those of the mind, what is the relation of *their* logic to logic itself, that is,
the logic of the mind—the logic of belief? Moreover, if our affects have
their own reasons and logic, what implications does this fact hold for our
understanding of reason itself? That is, what does it tell us about human
cognition if cognition is in some way able to accommodate the dynamics

1

of our affective lives? If there is a logic of the heart, what does it reveal about the processes at the heart of logic?

Defenders of the rationality of the emotions, proponents of cognitivist theories, have commenced their demonstrations from a familiar topos: the critique of a century-old, once-influential theory of the emotions, that of William James. Typically, James has been taken as an exemplar of the type of theorist the cognitivist wants to define himself or herself against. What had once recommended James's theory was that, in its stress on felt bodily states, it immediately connected with what appears as the essential thing about emotions—how they *feel*. However, the serious shortcoming of his somatically based approach, according to its cognitivist critics, is that it allows no space for the operation of cognitive influences on emotions. Emotional reactions are not, they have claimed, simple causal effects of worldly stimuli; they are effects of the world understood and evaluated in certain ways.

With this idea, the cognitivists could connect with another everyday belief, as entrenched as the belief that emotions are feelings—the belief that typically there are reasons for feeling the way we do. But the difficulty at the heart of emotion theory has been to marry these two equally obvious dimensions of emotional life, the affective and the cognitive, in some effective and plausible way. Thus the cognitivists chided James for his neglect of cognitivity, but a growing number of critics suggest that the price paid for the cognitivists' way of incorporating reason into emotion has been that its cognitive side had been secured at the expense of the affective. Not surprisingly, then, after a period of "high cognitivism" in emotion theory in the 1960s and 1970s the pendulum started to swing back in the direction of Jamesian somaticism.

These disputes within the world of emotion theory might be regarded as local and peripheral to more general considerations about the nature of mind, but this attitude would, I believe, be near-sighted. On investigation they can be seen as a microcosmic reflection of much more general issues about the mind and as condensing a tangle of interlacing claims and counterclaims about the mind–body relation, consciousness, "mental representation," the relation of the mind's software to its hardware, the status of qualia or the mind's phenomenal characteristics, and a host of other complex and puzzling issues that have increasingly occupied theoretical psychologists, philosophers, cognitive scientists, neuroscientists, and others in the last decades of the twentieth century. The problem of reconciling cognition and affect in emotion is not just one single problem encountered in thought about the mind, let alone a minor or peripheral one. In significant ways it sums up the defining problems of mind for the philosophers and psychologists of our age. The investigation of affect by necessity touches on many other aspects of mental life; and so, while affect serves as a guiding thread in this book, we will need to explore issues well beyond the realm of emotional life itself. As any comprehensive treatment of all such issues must be out of the question, what investigative strategy are we to adopt?

Luckily there is an untapped store of insights from past investigators to be retrieved. First of all, having been reconstructed as the physiologically reductionist inverse of modern cognitive theory, James has often been reduced to caricature and his valuable insights about the mind squandered. In Chapter 1 I offer a quick sketch of some of the more relevant general issues that have dominated thought about the mind in the twentieth century, and in Chapter 2 I turn to a long-overdue reassessment of James's theory of the emotions. James's position, I contend, is both more complex and more receptive to the interrelation of cognitive and somatic-affective factors than it is credited for. However, to fully appreciate James we must understand him against the background of a now largely neglected nineteenth-century tradition, a background that he shared with the other great philosopher-psychologist of his time, Sigmund Freud. In Chapter 3 I show how from similar beginnings as James's, which drew on the ideas of his contemporaries concerning the "parallelism" of mental and somatic orders, the conception of the "cerebral reflex" and the notion of evolutionary layering of "higher" over "lower" centers of the brain, Freud was able to enunciate more clearly than James a thesis linking the dynamics of affective states and unconscious mental processes. He could do this essentially because of his *rediscovery* of a coherent conception of the unconscious mind, a conception that can be traced back one hundred years to Kant.

We tend to think of figures such as James and Freud from somewhere downstream of them, and "whiggishly" regard them as early figures of the twentieth century, rather than as late ones of the nineteenth.[2] But an advantage of thinking of them in the latter way, as thinkers at the close of a century rather than as revolutionaries issuing in a new one, is that it may allow us to retrieve useful, but now ignored or forgotten, ways of thinking about the mind. This is especially applicable, I believe, for a strand of thought rarely acknowledged in modern philosophy of mind, the strand of post-Kantian German idealism represented by Fichte, Schelling, and Hegel.

When figures from the history of modern philosophy are invoked in discussions of the mind, it is usually the names of Descartes, Locke, and Hume, or less commonly of Thomas Reid, that are raised. More recently, Kant as a theorist of the mind has started to be taken seriously as a progenitor of the popular "functionalist" conception of the mind. But more often than not, the list stops there, and one can get the impression of a tradition of modern European philosophico-psychological speculation as passing directly from the seventeenth and eighteenth centuries to the present, despite the fact that it was clearly the nineteenth century in which the modern understanding of the function of the brain took shape. Clearly this nineteenth-century revolution in the neurosciences that James and Freud inherited was influenced in significant ways by post-Kantian idealism, in particular in the form of Schelling's "nature-philosophical" ideas.

In Chapter 4 I reach back to the origins of this psychological tradition in Kant, whose views of the mind I examine in the light of Patricia Kitcher's recent attempt to claim Kant for modern cognitive science. But Kitcher's Kant, I suggest, faces the same sorts of problems that confronted the historical Kant and that have returned to haunt the cognitive theory of emotions, particularly the relation between the mind's representational capacities and its consciousness, and specifically its self-consciousness. It was in the context of disputes in the wake of Kant that Fichte formulated his ideas about the necessary role within the functioning of the conscious, representational mind of *self-feeling*, which was essentially unconscious and nonrepresentational. In doing so, Fichte opened up a fruitful, but largely ignored, philosophical perspective on the complex interrelation of cognition and affect.

Figures such as Fichte are commonly discounted as being of potential value for contemporary thought about the mind because of their espousal of idealism; but exactly what idealism amounts to in the post-Kantian context is in general poorly understood. On closer examination, as I show in Chapter 5, Fichte's view of the individual mind is unintelligible without the assumption that it exists as embodied and located in the world. His peculiar idealism he described as a "real-idealism or an ideal-realism,"[3] an idealism that defined itself in opposition to what he considered to be "transcendent" realism or "dogmatism," rather than realism per se. The Fichtean idea of the interdependence of realistic and "transcendentally" idealistic perspectives with respect to the mind was then pursued by the early Schelling and allowed him to attempt to incorporate into his philosophical speculation the biological ideas about the body that were then taking shape within the contemporary medical sciences. In Chapter 6 I examine Schelling's development of Fichtean ideas and show how his attempted identification of realist and idealist ways of looking at the mind gave rise to pre-Freudian ideas about the nature of unconscious mental function grounded in feeling and subjected to its own peculiar logic.

In Chapter 7 I turn to the third of the post-Kantian idealists, Hegel, and argue that he in fact continued Schelling's investigation of a peculiar form of affectively charged thought that, like the Freudian "primary process," worked on principles of polarity and analogy, while attempting to resist what he saw as the inherently irrationalist dimensions of the trajectory that Schelling's thought had assumed. Hegel is, of course, best known for his ideas about the socially and culturally borne and historically developing "spirit" to which any individual belonged and in terms of which he or she had to be understood. But cultural dynamics for Hegel were also tied ultimately to the physiological processes of the body as the locus of affects and their expression, which articulated the individual to others within the processes of culture. In Hegel, therefore, we see even further aspects of the surprisingly contemporary-looking account of feeling, its embodiment and its relation to cognition that is common to the three major post-Kantians. In this respect there are illuminating parallels between Hegel's theory of affect and that found in the work of Silvan Tomkins, a psychologist of affect who resisted the excesses of cognitivism in the period of its triumph. In Tomkins's work, we see re-emerge something of what a post-Kantian psychology might look like in the later twentieth century.

However, even if James and Freud did in fact inherit certain ideas from this earlier age, both of them espoused the type of evolutionary naturalism opened up just prior to them by Charles Darwin. Does not this make the earlier work of the idealists only of historical interest? In Chapter 8 I turn to this more general question of how the thought of an earlier idealist movement might possibly be relevant to naturalistic thought about the mind at the end of the twentieth century.

Surprisingly, it may turn out that the acceptance of an evolutionary perspective on the mind alters the landscape of idealist psychology in quite minimal ways. Sophisticated evolutionary accounts about the development of language in our humanoid ancestors and its relation to earlier communicative systems can produce a picture that is uncannily close to the type of general account of the mind and its dependence on both the body and its externalized cultural scaffolding found in Hegel. Moreover, even in Darwin himself we find intimations of a conception of a logic of affect essentially homologous to the one I trace from the idealists to Freud.

1

Affect in Twentieth-Century Thought

Modern psychological theories of emotion are commonly regarded as starting with the work of William James. For James, the subjective "feeling" of an emotion was provided by bodily states and processes, primarily conceived as located "peripherally" within viscera, skeletal muscle, and skin. When noted, such processes are typically thought of as effects or expressions of their associated emotions. But this, James claimed, was wrong; such bodily states were, rather, constitutive of those emotions themselves, not effects or expressions of some purely mental feeling:

> If we fancy some strong emotion, and then try to abstract from our consciousness of it all the feelings of its characteristic bodily symptoms, we find we have nothing left behind, no "mind-stuff" out of which the emotion can be constituted, and that a cold and neutral state of intellectual perception is all that remains. . . . What kind of an emotion of fear would be left, if the feelings neither of quickened heart-beats nor of shallow breathing, neither of trembling lips nor of weakened limbs, neither of goose-flesh nor of visceral stirrings, were present, it is quite impossible to think. Can one fancy the state of rage and picture no ebullition of it in the chest, no flushing of the face, no dilatation of the nostrils, no clenching of the teeth, no impulse to vigorous action, but in their stead limp muscles, calm breathing, and a placid face? The present writer, for one, certainly cannot. The rage is as completely evaporated as the sensation of its so-called manifestations.[1]

But the theory that came to be known as the "James–Lange" theory was soon challenged. James's postulated physiological states were not, it was argued, sufficient to individuate emotions; some other "cognitive" factors were needed. At first this opposition took a negative tack concerning the insufficiency of the "physiological discriminators" of emotion, a charge mounted early in the century by the influential physiologist Walter Cannon who had concluded that physiological states alone could not constitute specific emotions since "the same visceral changes occur in the very different emotional states and in non-emotional states."[2] But with the development of a distinctively "cognitive" turn in psychology in the second half of the twentieth century this criticism could now be linked to a comprehensive account of what James's physiological theory lacked. James's view of the mind now seemed to belong to a very distant age.

The Cognitive Turn against Jamesian Somaticism

In some ways aspects of James's somatic account of emotional states could be seen to be generally compatible with that behavioristic approach to the mind that had become dominant in psychology in the first half of the century. Within philosophy this outlook was represented by Gilbert Ryle's *The Concept of Mind*,[3] among others, but by the late 1950s and 1960s behaviorism came to be challenged by an array of other materialist accounts of the mind that regarded behaviorism as too restrictive. Within philosophy of mind one of the first of these was the "identity theory" postulating a straightforward identity between mind and brain: simply put, the mind, it was claimed, *was* the brain. Whereas behaviorists had vetoed talk of mental states and processes, only admitting discussion of behavior, this materialist identity theory admitted talk of the mental by equating it with the physical, or more strictly, the neurophysiological. The first kind of "mentality" to be treated in this way was sensation, when the philosopher J.J.C. Smart identified the mental event of having or undergoing a particular sensation, being in pain, say, with a physical event, the firing of certain neuronal fibers, for example.[4] Importantly, in centering the analysis on mental and physical events, Smart avoided talking of sensations as some kind of inner or mental "things," an approach to sensation that had been common since the seventeenth century. Smart's identity theory itself does not seem antipathetic to the Jamesian outlook. To describe the sensations of emotion as identical with the patterns of innervation originating in the endogenous stimulation of afferent nerves from viscera, muscle, and so on, could be seen as just an alternative way of putting James's point (although as we will see, James himself rejected this form of materialist identification of mind and brain). But Smart's type of identity theory was soon to be challenged by a new group of approaches to the mind that were beginning to assert themselves in both psychology and philosophy during the 1960s and that had strongly negative implications for Jamesian theory of the emotions. This group of approaches marched under the banner of "Cognitive Science."

In its later versions, such as that put forward in David Armstrong's *A Materialist Theory of the Mind* (1968), the materialist identity theory sought to identify brain states with a much broader range of mental states than sensation, including cognitive states, such as propositional attitudes such as "belief."[5] Within behaviorism, talk of mental states such as the holding of a certain belief was simply conceived as reducible to talk about *behaving* in certain ways. (Or to put it another way, mental events were merely logical constructions out of actual or possible behavioral events.) In contrast, within Armstrong's "Central State Materialism," having a belief was identified with being in a certain neurological state, which explained why the person with that belief was disposed to behave in the ways that concerned the behaviorists. Soon, however, a somewhat different form of identity theory appeared, still holding to the more generally materialist project underlying both behaviorism and the thesis of mind–brain identity, but unhappy with both the very thin account of the mind presupposed in the former and the overly physiological identification of mind with the human biological organ of the latter. This philosophical view paralleled the "cognitive" revolution in psychology and the development of disciplines such as computer science and artificial intelligence, and it put to good use the recently invented electronic computer as a model for the mind. This new philosophical approach was *functionalism*.[6] While identity theory identified mind with brain per se, functionalism identified mind with certain types of organizational patterns, functional states, capable of being instantiated in the brain, but, importantly, also capable of alternate forms of instantiation, such as in a computer.

Behaviorism focused on behavior, which it attempted to explain as the "output" of processes resulting from some sensory "input." Identity theory allowed talk of mental states as the cause of this behavior but identified them with physiological states. Like central state materialism, functionalism allowed talk of mental states, individuated in terms of its causal relations to sensory "inputs," behavioral "outputs," and other mental states. Such states were thought of in terms of abstract patterns indifferent to the medium in which they were instantiated. Such patterns were to be understood in terms of the notions of flow and processing of information.

The appropriate model for this existed in the "Turing machine," a hypothetical computational device whose internal states were able to be specified in terms of a "machine table" of rules. Computers were Turing machines, or approximations to them, and so it was readily appreciated that a material device could be thought of as having "mental states" in the sense suggested by the theory. While the computational states of a computer were clearly physical states linked to each other causally, they could also be considered as "syntactic" concatenations of symbols that were implemented in those physical states and that could be conceived as linked to each other in logical relations. In some versions of this functionalist perspective, as found in the work of Jerry Fodor for example, functional states were considered as "representational" ("sentences" written in a "language of the mind" such that the bearer of such sentences could be regarded as having "propositional attitudes," the sorts of beliefs, desires, hopes, etc., which we in everyday life attribute to ourselves and others). Such an approach, it was thought, could capture what was distinct about the psychology of cognitive beings like ourselves.[7]

As mentioned, such developments occurred in the context of the burgeonning of disciplines such as artificial intelligence (AI) and computer science, which, together with existing disciplines such as philosophy, psychology, and linguistics, eventually formed the core of the new sprawling confederation of disciplines that has come to be known as "Cognitive Science."[8] Especially significant for such developments was the contestation that behaviorism had received in linguistic theory, where Noam Chomsky had used mathematical models to posit the existence of underlying configurations of rules (constituting what Chomsky termed "linguistic competence") on the basis of which verbal behavior or "performance" could be explained. Within the parallel disciplines of AI and computer science, similar ideas about the specification of abstract levels of cognitive representation, separable from both the actual algorithms with which they could be computed and the "hardware" within which these algorithms were implemented, were put forward by theorists such as David Marr, Allen Newell, and Herbert Simon. Another important development within this complex of approaches was that of Warren McCulloch and Walter Pitts, who suggested that networks of neurons within the brain could be thought of as instantiating the same sorts of mathematico-logical rule structures making up the functional states of Turing machines. This introduced the idea of "neural networks," neurologically instantiated functional structures, that challenged the earlier model of the configuration of reflex arcs underlying behaviorism (as well as Jamesian psychology) and developed into the idea of a "connectionist" cognitive architecture that challenged the more classical symbolic or language of thought approach.

This developing cognitive turn had, predictably, important effects on the way that emotion came to be thought of and studied, and in the early

1960s there appeared an experimental study by psychologists Stanley Schachter and Jerome E. Singer that eventually achieved almost classic status as a definite cognitivist demolition of the James–Lange theory.[9] In a series of experiments Schachter and Singer induced in subjects states of physiological arousal by the use of norepinephrine injections while simultaneously attempting to manipulate those subjects' thoughts in determinate ways. In particular, the experiments were designed to test whether the same underlying physiological states would be interpreted by those subjects (who were uninformed of the actual effects of the injections) as manifestations of different emotional states (anger or euphoria, for example) depending on the nature of the cognitions (beliefs, interpretations) induced in them by the experimenters. Indeed, the results obtained seemed to confirm the thesis that cognitive considerations of "interpretation" or "labelling" entered into the determination of emotional states. From that time on the idea of the cognitive constitution of affective states came to play an ever more central role in orthodox psychological thought.[10]

It was from about this time as well that philosophical theories of the emotions started to take a similarly cognitive tack.[11] Common opinion might have it that an affective state, an emotion, a mood, a state of mind, was basically some type of brute, subjectively experienced feeling, capable perhaps of Jamesian identification with a bodily state; received philosophical opinion swung toward viewing them as cognitive or intentional states. Within analytic philosophy this was typically expressed in the analysis of emotional states in terms of *propositional attitudes*, that is, those attitudes directed to propositional mental contents that were being claimed as the states modeled by functionalists such as Fodor. In such analyses, the subtle intertwinings of beliefs, desires, hopes, and so on within typical emotional states were carefully reconstructed. Consider, for example, the emotion of envy: for me to be envious of my next-door neighbor's new car, my envy must be more than some simple brute feeling or some particular configuration of somatic states. Rather, it must be structured in terms of certain propositional attitudes such as my belief that the car is his, my desire that it be mine, and so on. What at first seemed like a mere feeling or felt bodily state turns out to be a highly complex cognitive state involving conceptualizations and "attitudes" intentionally directed to actual (in the case of beliefs) or desirable states of affairs in the world. From the perspective of such a richly cognitive view of the mind, affect as viewed from the perspective of the James–Lange thesis seemed to be hopelessly dumb.

This sketch inevitably oversimplifies a very complex scene. Not all adherents of philosophical functionalism saw it as cohering with the type of analysis of mental states that utilized propositional attitudes. Conversely not all cognitive scientists think of the brain's computational processes as being about the manipulation of items with propositional or symbolic logical form. Some insisted that attributing propositional attitudes to others was just a heuristic device useful for explaining their behavior, others thought of such structures realistically as functionally encoded in neural tissue. Nevertheless, what had been provided was a model that showed how mind and body could be put together in ways that preserved the ways in which each had been thought about independently and that made talk of cognitive states acceptable without the worry of an appeal to some Cartesian mental substance. James's critique of Cartesian "mind-stuff" was affirmed, but without loss of the type of talk of a cognitive reality that that mind-stuff had underpinned. James might have been an important figure in the history of psychology and philosophy of mind, but, from the viewpoint of such a cognitive science, history was where he belonged.

Affectless Emotions

The cognitive turn in psychology and allied sciences seemed, in contrast to the behaviorist epoch, to be on the side of the distinctly human: it allowed researchers in the human sciences to be both scientists and humanists. The general optimism over "the mind's new science" might be seen as part of the overall context within which the Singer and Schachter studies could come to achieve almost classic status within psychological discussions of affect. Nevertheless, the canonical cognitive view of the emotions was not without its critics. One in particular was Silvan Tomkins, a psychologist whose approach to emotion, although it provided a role for cognitive factors, nevertheless supported a more Jamesian bodily based understanding of the differentiation of emotional states. Tomkins pointed to the irony of the fact that the new cognitive approach, despite its espousal of a type of "mentalism," seemed strangely resistant to the evidence of conscious experience. "Surely no one who has experienced joy at one time and rage at another time would suppose that these radically different feelings were really the same except for different 'interpretations' placed on similar 'arousals.' Only a science which had come to radically discount conscious experience would have taken such an explanation seriously."[12] Moreover, the experimental work of some of Tomkins's former students seemed to suggest that Cannon's assumptions about the impossibility of physiological individuation of particular emotions was simply empirically mistaken. Thus it was argued that the list of universal "basic emotions" posited by Tomkins could, if fact, be differentiated at a purely physiological level.[13] James, it seemed, was not entirely forgotten.

While the cognitive revolt against behaviorism had chastized its blindness to the mental and prided itself on reintroducing "mentalism" to the science of the mind, Tomkins's criticism highlighted the somewhat particular view of what characterized "mentality" to many cognitivists. In opposition to behaviorism, cognitivists insisted on the legitimacy of talk about "mental states," but by this they typically meant the ascription to subjects of representational states, often understood as symbolically represented propositional attitudes. And with this cognitivists often preserved many aspects of behaviorism. For Tomkins they preserved behaviorism's resistance to the relevance of the conscious manifestation of emotions, that is, how they *felt*.

This indifference to consciousness became quite explicit in functionalist analyses of emotions. Thus Aaron Sloman, for example, analyzed emotions "as states in which powerful motives respond to relevant beliefs by triggering mechanisms required by resource-limited intelligent systems," adding *"physiological changes need not be involved."*[14] Sloman's dismissal of physiology was not an appeal to the view that James had criticized, the idea of a consciously felt content *without* the participation of physiology; rather, it was a rejection of the relevance of feeling *at all*. This belief in the inessential character of physiological states and their associated feelings could thus lead Sloman to the idea of the possible ascription of emotions to robots.[15]

While Tomkins had been something of a lone voice in the 1960s and 1970s, other criticisms of the cognitive theory of the emotions, like that of Robert Zajonc, started to become prominent in the 1980s. Zajonc has been critical of those cognitive psychologists, who, in the tradition of Singer and Schachter, had stressed the essential role of an interpretative "appraisal" of the cognitive stimuli in the production of some response. One confusing aspect of this debate centers on what at first seems to be a reversal of the positions of cognitivists and noncognitivists on the relevance of consciousness. While appraisal theorists presupposed that such "appraisals" were by necessity consciously accessible to the subject, critics like Zajonc argued that "subliminally" perceived or "unconscious" stimuli could trigger emotional reactions quite independently of, and even in opposition to, the subject's conscious appraisals. Thus in a synoptic article of 1984, "On the Primacy of Affect," Zajonc summarized the types of empirical studies that suggested the relative independence of emotion from cognition, including experiments that suggested that conscious appraisal and affect were often uncorrelated and disjoint or that particular affective reactions could be established without the participation of conscious appraisal.[16]

On closer examination, however, it can be seen that the problem seems to reside in vagueness and ambiguities concerning the notion of *consciousness*, a concept that had been positively expelled from much psychological thought throughout the twentieth century. Philosopher Ned Block has pointed to the "hybrid" or even "mongrel" nature of the concept of consciousness, differentiating "phenomenal consciousness" from what he terms "access consciousness."[17] Phenomenal consciousness concerns the peculiar felt quality of some experience, what Thomas Nagel had designated as the "what it is like" to undergo that experience.[18] But access consciousness concerns the ability to integrate one's experience into what the philosopher Wilfrid Sellars had spoken of as the "space of reasons": In Block's terms, a mental state is access-conscious if it is "poised for use as a premiss in reasoning, . . . poised for rational control of action, and . . . poised for rational control of speech."[19]

With his distinction Block was concerned to point out fallacies that were commonly made in analyses of puzzling medical phenomena such as "blindsight," in which these two senses of consciousness were, he thought, misleadingly conflated. But it was clear that Block was making explicit and general a distinction that others had put in different terms. Thus Nicholas Humphrey, for example, had revived Thomas Reid's distinction between sensation and perception for much the same purpose.[20] And the distinction seems to be what is needed to dissolve the apparent paradox concerning the cognitivist account of emotion. Thus we might say that while Tomkins criticized cognitivism for its neglect of the phenomenal consciousness of emotions, their very affectivity, Zajonc was criticizing cognitivists' assumptions that one normally had "access consciousness" to one's emotional states and processes.[21] Interestingly, here access and phenomenal aspects of consciousness seemed to pull in different directions.

The Critique of Cognitivism in the Philosophy of Emotion: Functionalism, Attitudes, and Phenomenality

Block's use of the notion of phenomenal consciousness testifies to the extent to which this topic had penetrated discussions in the philosophy of mind in the 1980s and 1990s. A singularly key work here had been Thomas Nagel's "What Is It Like to Be a Bat?" of 1974, in which he had taken the postulated "phenomenological fact" of the "what it is like" of bat consciousness to exemplify something crucial about the nature of consciousness that, he thought, eluded attempts at "objective" or "third-person" forms of explanation. For a time Nagel-type considerations of phenomenality were engaged primarily within general debates over the possibility or impossibility of naturalistic explanations of consciousness. It was only later that they started appearing within the more familiar debates about the nature on mental content. Thus even in the late 1980s Daniel Dennett could note that "most of the major participants in the debates *about mental contents* . . . have been conspicuously silent on the topic of consciousness."[22] Gradually, however, the issue of phenomenality was brought into closer contact with theorizing about the nature of intentional content, as exemplified by works such as Michael Tye's *Ten Problems of Consciousness*[23] and Fred Dretske's *Naturalizing the Mind*,[24] both of which purported to provide representationalist theories of phenomenal consciousness.

Against this background of growing interest in the topic of phenomenal consciousness, it is not surprising that the same sort of disquiet that had led Tomkins to criticize psychological approaches to the emotions that ignored feeling and had led Nagel to criticize the neglect of phenomenality in the philosophy of mind eventually appeared with philosophical theories of the emotions. As one critic of the orthodox cognitive view has put it: "There was a time when philosophers had quite a lot to say about feeling and its relationship to emotion. . . . More recently, philosophical discussions of emotion have shifted to its more 'cognitive' dimensions. . . . Feelings have been left behind."[25]

The orthodox view in the philosophy of emotion had, as we have seen, adopted an approach relying largely on the appeal to "propositional attitudes," such as beliefs and desires, resulting in an orientation in general like that of appraisal theory of psychology. Here, too, phenomenal and access consciousness were seen to pull in different directions. Claire Armon-Jones has noted how far this orientation had tended to abstract from the phenomenological characteristics that feature so centrally in the everyday idea of emotions as "feelings."[26] Linked to this was the way that standard approaches had tended to skew considerations of affective states away from those apparently "objectless" *moods,* such as depression or joy, toward more differentiated and articulable "objectual" *emotions,* such as envy and guilt. Depression or joyfulness, she insisted, are commonly just that, apparently with nothing in particular, certainly nothing "propositional," that they are "about." Furthermore, even within the range of "objectual" emotions, the orthodox cognitive approach had tended to peripheralize a whole range of affective states, such as "irrational" emotions like phobias or even aesthetic feelings, as these states do not seem to allow analysis in terms of beliefs. If suffering from a phobic fear of snakes, for example, I might experience terror by simply being presented with a photograph of a snake or even by actively thinking of a snake, and in neither case am I likely to believe that I am in any real danger.[27] In short, the strong focus on "access consciousness" here, too, seemed to go hand in hand with a relative indifference to issues of the phenomenal consciousness, or affectivity of emotions.

As Armon-Jones points out, in the 1980s an emerging group of "neo-cognitivist" critics of the orthodox philosophical view started to look to some alternative way of articulating the cognitive component of emotions, invoking ideas such as "non-propositional sensing," "aspect perception," "seeing as," or "construing."[28] Such an appeal to nonpropositional forms of representation adequate for the treatment of emotions has also been a feature of the work of theorists, who, working from perspectives more aligned with cognitive science, have also been concerned with the issue of phenomenality. For example, both Louis Charland and Michael Tye assert that affective states, while "representational," do not represent in the way that beliefs do: they are "non-" or "infradoxastic."[29]

In his representational theory of phenomenal consciousness, Tye lists perceptual experiences, bodily sensations, and affective states such as passions, emotions, and moods as states that are typically phenomenally conscious and excludes propositional attitudes like desires and beliefs. "It seems to me not implausible to deal with these cases by arguing that insofar as there is any phenomenal or immediately experienced felt quality to the above states, this is due to their being accompanied by sensations or images or feelings that are the real bearers of the phenomenal character. Take away the feelings and experiences that happen to be associated with the above states in particular cases, and there is no phenomenal consciousness left."[30] A similar position on the nonphenomenality of propositional attitudes has also been advanced by Norton Nelkin who sees the traditional tendency to model mental states on the paradigm of sensation (a lingering effect of early British empiricism, especially that of Hume) as behind the assumption that attitudes will be conscious in the phenomenal, Nagelian, sense. But for Nelkin, as for Humphrey, it was Reid rather than Hume who got the relation right: "Attitudinal-consciousness is an altogether different sort of consciousness from sensation-consciousness."[31] While sensations are the mental states that are characterized by phenomenality, "In Nagel's sense there is nothing that it is like to believe consciously."[32] And if something like this distinction is right, it is clear why the standard cognitivist view of emotions has looked like it left out their felt aspect—their very affectivity. By restricting its analysis of mental contents to beliefs and desires, it had focused on mental contents that cannot have phenomenal characteristics.

Bringing Back the Body: The Biology of Emotion

With such developments, the strongly cognitivist attitude toward affect that had earlier appeared to have consigned James's theory to history has come to look increasingly shaky. And while the sort of functionalist or computationalist perspective to which cognitivism was commonly linked had prided itself on its *naturalism,* an approach to emotion that can intelligibly attribute it to machines but not babies might be regarded as a questionable form of naturalism indeed. Thus it is not surprising that the proponents of a naturalistic understanding of emotions would have tacked back toward the Jamesian focus on the body.

In the mid-1990s two popular and synoptic books by leading researchers on the biological basis of emotion added to the critique of strong cognitivism, *Descartes' Error: Emotion, Reason, and the Human Brain* by the neurologist Antonio R. Damasio,[33] and *The Emotional Brain: The Mysterious Underpinnings of Emotional Life* by Joseph LeDoux,[34] an experimental physiologist. Both these books, while serving to popularize the basic theoretical, or experimental/clinical approach of their authors, provided synoptic overviews of current biological and evolutionary perspectives on emotion. A striking feature of these books was the support they offered for the Jamesian thesis that the felt-center of emotion consists of informational feedback from states of the body. Another, given their convergence with accounts such as Zajonc's, which discussed the (access) unconscious aspects of emotion, was the indirect support they provided for some of the basic tenets of another late-nineteenth-century, early-twentieth-century psychologist, Sigmund Freud. Curiously, both books have the feel of the late-nineteenth-century *Zeitgeist,* a naturalistic outlook strongly based in evolutionary biology but willing to talk about consciousness and subjectivity in a way that has been excluded from most of the twentieth century.

For LeDoux, "the response of the body is an integral part of the overall emotion process. As William James, the father of American psychology, once noted, it is difficult to imagine emotions in the absence of their bodily expressions."[35] The processes of cognition, as the functionalists argue, may be able to be understood in abstraction from their physiological embodiment, but not so emotion: "[Emotions] evolved as behavioral and physiological specializations, bodily responses controlled by the brain, that allowed ancestral organisms to survive in hostile environments and procreate. If the biological machine of emotion, but not cognition, crucially includes the body, then the kind of machine that is needed to run emotion is different from the kind needed to run cognition."[36]

LeDoux bases his acceptance of Jamesian feedback theory on a number of facts: phenomenologically, we *feel* our emotions in our body in this way; the contestable nature of much of the empirical evidence that had been, since Cannon, advanced *against* James; the existence of ample neurological feedback from the body during emotional experience; and the experimental evidence that subjective states of emotion can be induced artificially by bringing about particular somatic changes, especially to the musculature of the face.[37] For his part, Damasio, on the basis of neurological deficit studies, advances a particular version of the somatic feedback theory he calls the "somatic-marker hypothesis," proposing a physiological basis for the role of "gut feeling," not only in the experience of emotion, but also in practical decision making.[38]

Influenced by such types of studies Daniel Dennett, a strong advocate of functionalist accounts of the mind and consciousness, has also incorporated biological considerations into his theory of consciousness in a way that seems readily applicable to emotion. He points out that many cognitive scientists had implicitly accepted a type of neo-Cartesian duality between the brain as media-neutral information processor and body as the site of "transducers" and "effectors" (or "input" and "output" nodes). Transducers are those mechanisms which translate information from one medium to another (both photoelectric cells and the rods and cones of the retina transduce light into their own particular types of signals), while an effector is "any device that can be directed, by some signal in some medium, to make something happen in another 'medium' (to bend an arm, close a pore, secrete a fluid, make a noise)."[39] But, continues Dennett, while the control systems of human-made machines (the steering mechanism of a ship, for example) are typically isolated from that which is controlled in this way, such a separation does *not* apply to the biological "control systems" of animals, and, by implication, of ourselves,[40] "because they evolved as the control systems of organisms that already were lavishly equipped with highly distributed control systems, and the new systems had to be built on top of, and in deep collaboration with, these earlier systems, creating an astronomically high number of points of transduction."[41]

Given its evolutionary history (e.g., transducers and effectors are implicated within information transmitting and processing operations), the human nervous system is not a "pure" signaling system that transmits information but almost no energy, and this, Dennett seems to imply, is somehow implicated in the fact of our phenomenal consciousness.[42] Dennett has been a long-standing critic of the view that either sentience or intentionality results from some *intrinsic* properties of neural tissue. But when he notes that "the neural control systems for animals . . . are not really media-neutral—not because the control systems have to be made of particular materials in order to generate that special aura or buzz or whatever,"[43] I assume that with this Dennett means that there *is* actual "aura" or "buzz" (sentience, or phenomenal consciousness), but that, rather than resulting from anything special about the particular materials from which the brain is built, it results from the fact that our brains were built on top of an earlier system, with the resulting mixture of the processes of information processing on the one hand and transducing on the other.

Body, Affect, and Representation

Daniel Dennett is no friend of "qualia," those postulated phenomenal characteristics of experience, and I have, somewhat liberally, interpreted his words as sketching a theory of phenomenal consciousness. In an article on the nature of affect, however, Louis Charland does not share Dennett's reluctance. For him: "The feedback hypothesis provides an interesting model for explaining what the qualia in emotion are and how they are generated. Qualia are transduced interoceptive representational feedback from internal physiological centres."[44] Charland insists that "feelings are representational . . . feeling *is* representing," in fact, for him affect is a form of perception, specifically an "interoceptive" form which is "designed to pick up and processes affective information about an organism's inner physiological states and processes."[45]

A similar view is found in Michael Tye's account of affective states. "Simple felt moods and emotions are sensory representations similar in their intentional character to background feelings and bodily sensations like pain." In anger, say, various physiological changes are produced and "the feeling you undergo consists in the complex sensory representations of these changes."[46] In contrast to the orthodox cognitivists we have seen, Tye does not hold that emotions and moods represent in the way that beliefs do. Rather, such sensory representations are natural representations of what they represent, in the way that the number of rings in the cross section of a tree trunk represent the age of the tree. Natural representation is a matter of objective facts, those of "causal covariation or correlation (*tracking*, as I shall often call it) under optimal conditions. If there are no distorting factors, no anomalies or abnormalities, the number of tree rings tracks age, the height of the mercury column tracks temperature, the position of the speedometer needle tracks speed, and so on. Thereby, age, temperature, speed, and the like, are represented."[47]

One might ask, however, exactly why causal covariation or correlation should be taken as counting as *representation*. It is clear that such things can be *taken* as representations in the sense of indicating or containing information about something, but in this sense there is someone indicated to, or informed. But this is not what Tye seems to mean, as he describes such representation as "observer independent": "Before any human ever noticed any rings inside trees, the number of rings represented the age of the tree."[48] But is this a meaningful use of the word *represent?*

In his account of representation Fred Dretske has also appealed to the idea of "nomic dependency" in defining representation, an idea that is essentially the same as Tye's "tracking."[49] As an event occurring in the thermometer, a change in the height of the mercury, stands in a lawlike relation to (in Tye's terms "tracks") an event occurring in the immediate environment of the thermometer, a change in its temperature. The height of mercury in the thermometer can be said to carry information about, or to "represent," the ambient temperature. But for Dretske, while thermometers, galvanometers, and so forth can carry information, they cannot be thought of as having "cognitive states," albeit even extremely primitive and simple ones. This is not because they fail to have information or intentional content. Quite the opposite—it is rather a problem of an abundance of such information. Instruments such as thermometers and galvanometers "cannot distinguish between pieces of information that, from a cognitive standpoint, are different." The position of the needle of a galvanometer, for example, will carry information not only about the amount of current flowing between its terminals, but also about the voltage difference between its terminals, as it will about the tension within its spring and a wide variety of other factors. As information has been defined in terms of lawlike relationships, information about F will always be also information about G if F and G are related in a lawlike way. But knowledge is not like that. Knowing one piece of information does not imply that the knower knows all "nomically equivalent" pieces of information:

> If there is a natural law to the effect that every F is G, then no information-processing system (man included) can occupy a state having the informational content that something is F without, thereby, occupying a state (the same state) having the informational content that something is G. . . . Therefore, if a system is to be capable of knowing that x is F without knowing that x is G . . . this system must be endowed with the resources for representing nomically related states of affairs in different ways. . . . For this to occur, the *cognitive* content of the system's internal states must be a function, not only of the information they are designed (or were evolved) to carry, but of *the manner* in which this information is represented or coded.[50]

Not all representations are, of course mental representations, but sensations or feelings are certainly mental. It would seem that for Dretskean reasons affective or sensory states if representational could not be so on the basis of "nomic dependency" alone.[51] This issue of how to understand the relation of the representational aspects of the mind to its sentience will recur throughout this book. What I wish to do for the time being is to raise the question of the impulse to bring, as does Tye, the language of "representation" to states that, such as affective states, have phenomenal qualities. Compare Tye on sensation to Smart, for example.[52] Smart does not talk of sensation as representing some particular bodily event with which it is correlated; rather, sensation *is* that event. Alternatively, as we will see, a popular view in the nineteenth century talked of the "parallelism" between sensory states and their correlated neurophysiological states.[53] What then is at issue in the desire to talk of phenomenal states as representational?

Perhaps we can see in Tye's account something of the peculiar "Cartesianism" of brain–body dualism that Dennett has alluded to as prevalent in much cognitive science. Tye wants to dismiss the idea that his notion of natural representation needs anyone *for* whom the representation represents. But in fact there does seem to be an implicit something "for" which the sensation is a representation in Tye's account: that something is the brain. But even though sensation might be the brain's way of "knowing" what is going on in the rest of the body for the purposes of its regulatory functions, qua phenomenal states it is not my brain that has sensations, it is *me;* and with this relation it is not at all clear that the language of information and representation is relevant. After all, why do I need to be informed about my body: it is not as if it is somewhere else, such that I need to be "sent information" about it. How closer can one get than one is to oneself?

To raise this question however brings into focus the fact that I *can* orient myself epistemically to the states of my body, as when I direct my attention to some particular sensation. But this ability itself seems to presuppose some prior, less epistemic, contact. "Does it just feel cold when I dab this on your tooth?" asks my dentist, "Or would you describe it more as a slight pain?" I suddenly shift all of my attention to some aspect of my experience I had been desperately trying to ignore. In such cases, it does seems to make sense to talk about the state that I can come to be in as a "representation" and as if it has some sort of epistemic content. But was the sensation a representation *before* I so directed my attention? And it is important that this question is not merely about how large the sensation looms within one's consciousness. Consider the difference between the straightforwardly interoceptive stance I can assume toward some sensation and a situation, such as may be found say, in intense pains or pleasures, where one is just in that state, "lost" or "absorbed" in it, as it were.

Ned Block has appealed to orgasm as an example of a clearly nonrepresentational sensation, and there does seem to be something intuitively right about this.[54] One might, if one were conducting an experimental study of sexual phenomenology, adopt an interoceptive attitude to one's orgasmic states while in them, but this is clearly different to, and not nearly as much fun as, what we normally think of as *having* an orgasm.[55]

The distinction between being in a phenomenal state such as orgasm or pain and consciously representing one's body by that state might also be seen as posing a problem for the "Reidian" position of Humphrey, who distinguishes between the paradigmatically representational perception and the paradigmatically phenomenal *sensation*, and yet at the same time tends to equate the latter with a *different* type of representational state, that of the "autocentric" (as opposed to "allocentric") representation—a representation which "provides the answer to the question 'What is happening to me?' "[56]

Humphrey uses Reid's distinction in interpreting the phenomenon described by Paul Bach-y-Rita in terms of the concept of "substitute sensation." Exploring ways in which the blind may be assisted by a type of technologically enabled "vision," Bach-y-Rita had fitted blind subjects with small video cameras the outputs of which were fed to "screens" each consisting of a matrix of 400 tiny vibrators. In each case the screen was applied to a patch of skin on the subject's body, the idea being that with this crude tactile image projected onto that area of skin, that skin might be able to function as a type of substitute retina, allowing these patients to "see." The experiments were indeed successful, the blind subjects after a period of training developing the capacity to recognize objects in space, judge distances and the size of objects "seen" and so on.[57]

Humphrey is interested in the question of whether such tactile sensations, when incorporated in the task of quasi-visual imaging, continue to be felt *as* sensations, that is, as something "happening to me." Bach-y-Rita had reported that even during the "visualizing" task a subject could feel the tactile sensations as such, but that "unless specifically asked, experienced subjects are not attending to the sensation of stimulation on the skin of their back, although this can be recalled and experienced in retrospect."[58] The sensation, it would seem, could be interpreted *either* as a sensation at the site of the stimulus, *or* as a perception, a representation of an external object.

If we understand the "sensation" *as* a representation (of the body) we would seem to need to appeal to a *third* state, that state which can occupy the two different functional roles in interoception and exteroception and stands to the representational sensation, just like the orgasm *had* stands to the orgasm felt *as* a representation of bodily states. This, then, would suggest that Reid's binary distinction needs to replaced by a three-way one: that between being in sensational states and representing by means of them, either extero- or interoceptively. In later chapters I argue that such phenomena point to a schema closer to that found in post-Kantians such as Fichte, Schelling, and Hegel, where sensation or "feeling" is conceived as phenomenal but "unconscious" in the sense of lacking "access consciousness." Such a schema might help us to appreciate the work of James and Freud, two theorists of affect to whom I now turn in the following chapters.

2

James's Theory of the Emotions
in the Context of His Conception
of the Mind

In their experiments Schachter and Singer were in fact building on a long tradition of experimental testing of the James–Lange theory of the emotions. As mentioned, the criticism of a lack of "clear-cut physiological discriminators of the various emotions" had been made from the basis of experimental studies that had culminated in an influential work of the physiologist Walter B. Cannon. Significantly, however, the theory tested and criticized as the James–Lange theory by Cannon and later Schachter and Singer seems quite a different one from the one actually found in James's original works.

James's Actual Theory of Emotion

Cannon had summarized James's theory "in nearly his own terms" thus: "An object stimulates one or more sense organs; afferent impulses pass to the cortex and the object is perceived; thereupon currents run down to muscles and viscera and alter them in complex ways; afferent impulses from these disturbed organs course back to the cortex and when there perceived transform the 'object-simply-apprehended' to the 'object-emotionally-felt'."[1] This description is in fact a close paraphrase of a passage toward the end of James's 1884 article "What Is an Emotion?"[2] Isolated from the rest of the article and from other of James's writings, however, it seriously distorts James's theory.

Crucially, Cannon's summary statement makes James sound as if he held a somewhat Lockean atomistic and associationist conception of perception underwritten by the mechanism of the reflex arc. Thus, an object is first "simply-apprehended" via the passive reception of impulses from the senses, next, and consequent upon this, discharges pass to the muscles and viscera altering their states, and finally, afferent discharges from the muscles and viscera running back to the brain result in a new set of endogenously originating perceptions that become associated with the original external perceptions to produce the "object-emotionally-felt."

This picture is, however, misleading. As is obvious earlier in the article and elsewhere, James had what might be described as a much more "Kantian" conception of perception (Kantian inasmuch as he had stressed the active contribution of the perceiver to the perceptual process itself and to its representational outcome).[3] This was, however, a *somatic* Kantianism, a Kantianism given a naturalistic twist by the growth of perceptual physiology on the one hand and evolutionary theory on the other, a view perhaps shaped by Herbert Spencer's earlier "evolutionary Kantianism."[4] Within James's somatic Kantianism, what can count as a "stimulus" is itself immediately determined by something about the organism's sensory-motor makeup, but given that this had coevolved with its environmentally directed motor capacities, the possible stimulus is further determined by those very resources that will produce its appropriate response. Thus: "As surely as the hermit-crab's abdomen presupposes the existence of empty whelk-shells somewhere to be found, so surely do the hound's olfactories imply the existence, on the one hand, of deer's or foxes' feet, and on the other, the tendency to follow up their tracks."[5] In short, the hound's experiential apparatus is no *tabula rasa* but rather is preset to register certain specific worldly things, "deer's or foxes' feet," and the nature of this presetting is itself linked to the hound's typical form of action, tracking. This means that "the neural machinery is but a hyphen between determinate arrangements of matter outside the body and determinate impulses to inhibition or discharge within its organs."[6]

This holistic idea that what can count as a stimulus is already determined by the constitution of the organism itself, while clearly present in James's understanding of the reflex arc, was to be made even more explicit in an article by John Dewey in 1896:

What we have is a circuit, not an arc or broken segment of a circle. This circuit is more truly termed organic than reflex, because the motor response determines the stimulus, just as truly as sensory stimulus determines movement. Indeed, the movement is only for the sake of determining the stimulus, of fixing what kind of a stimulus it is, of interpreting it.[7]

This idea has been seen as anticipating the later development of the "cybernetic" concept of the feedback loop, but, as we will see, it is an idea that can be traced back to early-nineteenth-century developments in the neurosciences, developments linked up with the often misunderstood "romantic nature-philosophy" movement in early-nineteenth-century biology. In the present context, however, it is in the light of Darwinian evolutionary theory, and in particular, Darwin's own theory of the emotions in *The Expression of the Emotions in Man and Animals,* that James argues for his conception of the nature of the reflex and the understanding of the nature of emotion consequent upon that conception.[8]

In "What Is an Emotion?" James anticipates an objection to his idea of the reflex as evolutionary mechanism serving the "connate adaption of the nervous system to that object" by objecting that surely "most of the objects of civilized men's emotions are things to which it would be preposterous to suppose their nervous systems connately adapted." Specifically, it is the conventionality of the things that occasion emotion in "civilized men." Would not this mean, then, that here bodily changes follow ideas; and "if in *these* cases the bodily changes follow the ideas, instead of giving rise to them, why not then in all cases?"[9]

We can see in this anticipated objection something of the objection later put forward by Schachter and Singer about the determining role of "cognitive" factors such as ideas, beliefs, interpretations. James's reply here is brief but crucial. First, he recalls "the well-known evolutionary principle that wherein a certain power has once been fixed in an animal by virtue of its utility in presence of certain features of the environment, it may turn out to be useful in presence of other features of the environment that had originally nothing to do with either producing or preserving it. A nervous tendency to discharge being once there, all sorts of unforeseen things may pull the trigger and let loose the effects. That among these things should be conventionalities of man's contriving is a matter of no psychological consequence whatever." Then noting that "the most important part of my environment is my fellow-man," and that "consciousness of his attitude towards me is the perception that normally unlocks most of my shames and indignations and fears," James expands on the notion of one's "consciousness" or "perception" of the attitude of another person.[10]

> It is not surprising that the additional persuasion that my fellow-man's attitude means either well or ill for me, should awaken stronger emotions still. In primitive societies "Well" may mean handing me a piece of beef, and "Ill" may mean aiming a blow at my skull. In our "cultured age," "Ill" may mean cutting me in the street, and "Well," giving me an honorary degree.

What the action itself may be is quite insignificant, so long as I can perceive in it intent or *animus*."[11]

This passage surely constitutes compelling evidence against the idea that James is postulating physiological as distinct from cognitive determinants of emotion. It is clear that here the word *perceive* is meant quite literally: one *perceives* another's intention. That is, here, at least, what James means by perception is *interpretative* perception—a perceiving *as*.[12] And, moreover, as the passage makes out, it is this type of interpretative perception that "awaken[s] . . . emotions." In short, for James, cognitive factors *do* enter into the determination of emotions; moreover, they enter into the determination of emotions in that way advocated by the recent neocognitive school, via an interpretative perception that construes its object as some definite type of object. Such a cognitive story James seems to consider as compatible with, indeed, as part of, the physiological story he tells, and not opposed to it.[13] In fact, a very significant difference in how James and his critics view the relation between psychology and physiology is revealed in their respective understandings of the reflex arc implicit in the perceptual triggering of an emotion, a difference rooted in nineteenth-century neuroscience.

Cognition and Mechanism: The Cerebral Reflex

Although elements of the modern neuroscientific concept of reflex action had developed since the late eighteenth century, it was the work of the British neurologist Marshall Hall (1790–1857) from the 1830s that gave rise to the modern understanding of the reflex arc and that moved it "away from the nebulous notions of soul and other immaterial principles and toward an explanation . . . based on anatomy and the unique functions of the spinal cord."[14]

Within a short time, however, Hall's limitation of reflex action to the spinal cord and the medulla oblongata (lower brain stem) was criticized by researchers, especially in Germany, who wanted to extend the idea of reflex action to the "highest" parts of the brain with the idea of a "cerebral reflex." As Clarke and Jacyna have argued in their work on the history of modern neuroscientific thought, this dispute over the notion of a cerebral reflex reflected different underlying metaphysical doctrines. Hall, together with most of his British contemporaries "continued to uphold the doctrine that the soul could act upon the cerebrum to produce actions, and . . . had no intention of superseding this spiritualist model with a theory of cerebral reflexes."[15] That is, the higher cognitive functions were conceived as working on principles entirely different from those implicit in the neurological reflex. But such "conventional dualist assumptions" were criticized by a variety of researchers in Germany where earlier the influence of Schellingian *Naturphilosophie* had led to the acceptance of a much more unified account of mind and body. "In an intellectual setting where the rigid Cartesian mind–body dichotomy had been greatly eroded, it was less plausible to consider the reflex as something entirely divorced from mind; rather, it was seen as yet another modality by which the soul ensured the integrity and continuance of the organism."[16]

In England, early exceptional adherents to the more German anti-Cartesian view of the mind–brain relation were Thomas Laycock (1812–1876) and William B. Carpenter (1813–1885), who both belonged to a circle of London-based, often Edinburgh-educated physicians and scientists who in the 1830s and 1840s became strongly influenced by German idealism and nature-philosophy.[17] Influenced especially by Carpenter, Herbert Spencer popularized the notion of the cerebral reflex within evolutionary circles. Later in the century, it was applied to explain symptoms of aphasia by Laycock's former pupil, the philosophically inclined London neurologist John Hughlings Jackson (1835–1911).[18]

This was a tradition with which James was well familiar. In *The Principles of Psychology*, while mentioning Jackson, James develops his account of reflex action on the basis of models provided by the Viennese cerebral anatomist and psychiatrist Theodor Meynert (1833–1892) in his classic work of 1884, *Psychiatry: A Clinical Treatise on Diseases of the Fore-Brain Based upon a Study of Its Structure, Functions, and Nutrition*.[19] In the course of his discussion of the "Meynert scheme," however, James introduces his own qualifications to counter what he describes as Meynert's excessively mechanistic understanding of the peripheral reflex, which "probably makes the lower centers too machine-like and the hemispheres not quite machine-like enough."[20] This latter point has a clear reference to the idea of cerebral reflex, and by the concluding section to this chapter, James has tentatively accepted the Jacksonian "cerebral reflex" in contrast to the Meynert scheme as the basis for higher cognitive functioning:

> Wider and completer observations show us both that the lower centers are more spontaneous, and that the hemispheres are more automatic, than the Meynert scheme allows.[21]

> Such cerebral reflexes, if they exist, form a basis quite as good as that which the Meynert scheme offers, for the acquisition of memories and associations which may later result in all sorts of "changes of partners" in the psychic

world. . . . We thus get whatever psychological truth the Meynert scheme possesses without entangling ourselves on a dubious anatomy and physiology.[22]

Later, the reflex operation of higher processes is discussed with respect to the will, habit, instinct, and emotion.[23]

In the context of his work on aphasia, which was to directly influence the thought of Sigmund Freud, Jackson had contested the form of associationism that Meynert espoused.[24] Meynert had understood the link between sensory input and reflex action as being regulated by descending neural tracts from the cortex. The cortex itself was thought to receive direct neural pathways from the sensory surfaces of the body so that these surfaces were mapped in a topographical point-by-point manner. With experience associations were laid down between these cerebral points corresponding to a classical "association of ideas." Conversely, fixed topographical projections ran from the motor cortex to muscle groups. Significantly, such a conception of neural architecture had led Meynert to explicitly reject Darwin's conception of emotional reflexes, arguing in ways that anticipated James's later cognitivist critics: the expression of emotions, such as the baring of the teeth in anger, were necessarily controlled by the cortex and could not be understood as reflex actions.[25]

In contrast, in the cerebral reflex scheme adopted by James cortical representation was viewed as mediated by lower reflex centers. Jackson, for example, discussed such an ascending hierarchy of reflex centers as involving the "re-representation" (and "re-re-representation"!) of the sensory and motor periphery.[26] This allowed him to assert, in contrast to Meynert, that it was particular goal-directed *functions* that were represented neurally: "nervous centres represent movements, not muscles."[27]

For James it was such a cognitive and interpretative dimension, a dimension to be understood functionally in terms of the organism's actions in an environment, that accrues to perception when considered as the afferent pathway of a unified and "educable" cortical reflex. And it was precisely this dimension that was subsequently left out of Cannon's purely physiological and mechanistic rendering of James's theory, and ignored by Schachter and Singer in their demand that James's theory of the emotions be supplemented by "cognitive" factors. In these later critical accounts James is reinterpreted as a type of reductive materialist or proto-behaviorist, for whom "subjective" or "cognitive" features such as beliefs or interpretations are eliminated within or reduced to the impersonal third-person discourse of physiology. But from James's point of view such eliminationist or reductionistic approaches actually presupposed the dualism that he had strenuously denied: for him any "purely" physiological conception of the nervous system without reference to the subjective aspect of these processes was just an "unreal abstraction":

> I hope that the reader will take no umbrage at my so mixing the physical and mental, and talking of reflex acts and hemispheres and reminiscences in the same breath, as if they were homogeneous quantities and factors of one causal chain. I have done so deliberately; for although I admit that from the radically physical point of view it is easy to conceive of the chain of events amongst the cells and fibres as complete in itself, and that whilst so conceiving it one need make no mention of ideas, I yet suspect that point of view of being an unreal abstraction. Reflexes in centres may take place even where accompanying feelings or ideas guide them.[28]

It was this refusal to eliminate the subjective and cognitive dimension from his more somatic analyses that could allow James to write in the chapter on Emotion in *Principles:*

> Let not this view be called materialistic. It is neither more nor less materialistic than any other view which says that our emotions are conditioned by nervous processes. No reader of this book is likely to rebel against such a saying so long as it is expressed in general terms; and if any one still finds materialism in the thesis now defended, that must be because of the special processes invoked. They are *sensational* processes, processes due to inward currents set up by physical happenings. Such processes have, it is true, always been regarded by the platonizers in psychology as having something peculiarly base about them. But our emotions must always be *inwardly* what they are, whatever be the physiological ground of their apparition.[29]

The materialism that James is opposing here then seems to be one that, in the manner of physicalists in recent times, aims at eliminating the reality of the "inner" processes by reductively explaining them in terms of mechanistically understood physiological processes. There are, according to James, some, the "platonizers," who would see his theory as materialistic because of its stress on sensational processes over that of higher-level ideation. But the implication of this is that James does not accept such a presupposition as the basis of deeming his theory to be materialist because it is bound up with an inadequate understanding of the mental. Again we find Dewey neatly capturing these Jamesian points in his article of the reflex arc: "We ought to be able to see that the ordinary conception of the reflex arc theory, instead of being a case of plain science, is a survival of the metaphysical dualism, first formulated by Plato, according to which the sensation is an ambiguous dweller on the border land of soul and body, the idea (or central process) is purely psychical, and the act (or movement) purely physical."[30]

James and Psychophysical Parallelism

The neglected aspects of James's thoughts about emotion that I have developed, aspects coherent with his strongly anti-Cartesian stance, suggest that we should see him as subscribing to an underlying account of the mind–body relationship that is radically at odds with the more dualistic accounts that treat "cognitive" processes as analyzable in abstraction from somatic ones. At first approximation, it suggests that James subscribed to some version of the type of view about the mind–brain relation held by the neurologically oriented researcher Hughlings Jackson and a variety of other biomedical researchers in the nineteenth century. This view is commonly referred to as the "dual aspect" theory or "psychophysical parallelism."[31]

This view, commonly related back to the philosophy of Spinoza, in which mind and extended matter were seen as different "aspects" of the one underlying substance, had been a commonplace among the more nature-philosophically inclined German medical thinkers of the early part of the nineteenth century. Thus Johann Christian Reil, an important German anatomist at the turn of the nineteenth century, whose support of the "paradigm change" resulting in the modern view of the cerebrospinal system seems to have been influenced by the nature-philosophy of Schelling, asserted in 1804 that "the real and the ideal sides of man are manifestations of a single being and organism that develops in two directions and thereby establishes the close dependence of the one on the other. In this case there would have to be a higher theory of nature to which physiology and psychology are subordinated in as much as the former deals with the objective side and the latter with the subjective side of that single fundamental basis of nature."[32] In Germany in the second half of the century, despite the general eclipse of nature-philosophy and the turn within physiology to the more reductionistic "biophysics program" of the 1840s and 1950s, the notion was carried forward into the new discipline of psychophysics by its founder, the Leipzig physicist G. T. Fechner.[33]

Fechner attempted to capture the relation of psychological to physiological as opposed aspects of the one reality by an analogy later commonly repeated: the subjective mind and the objective body should be grasped as if they were two sides of the one curve, the curve of a circle that, when viewed from a perspective centered *inside* the circle was convex and when viewed from a perspective centered outside it, was concave.[34] "Both sides belong together as indivisibly as do the mental and material sides of man and can be looked upon as analogous to his inner and outer sides. . . . What will appear to you as your mind from the internal standpoint, where you yourself are this mind, will, on the other hand, appear from the outside point of view as the material basis of this mind."[35] In fact, Fechner had been converted to nature-philosophy by reading the work of Lorenz Oken, a "transcendental anatomist" and early popularizer of Schellingian views in comparative anatomy. Schelling, who at the turn of the century had attempted to combine nature-philosophical ideas with Kantian transcendental idealism, had indeed enjoyed much influence within German medical and early psychiatric circles, and psychophysical parallelism provided a view that fitted neatly with the doctrine of the "cerebral reflex" in virtue of its opposition to traditional Cartesian dualism.

In Britain parallelist ideas took hold in the more idealist-influenced medical writers but became influential particularly through the physiological and psychological writings of George Henry Lewes. Lewes was well-versed in German philosophy and high culture and was an enthusiastic follower of Goethe's literary and scientific endeavors. Although Lewes espoused a Comtean positivism from the 1840s, according to his biographer he "went on reading and rereading both Kant and Hegel especially in the 1870s during his preparation for the *Problems of Life and Mind*, and he corresponded with the leading teachers of philosophy in Britain and Germany."[36] Herbert Spencer also adopted a type of psychophysical parallelism in his "aestho-physiology."[37] It was in this ambience that the philosophical-inclined Jackson, who had at one stage considered leaving neurology to pursue philosophy, from the 1870s utilized such a framework in an attempt to understand the psychological impairments suffered by patients with various forms of neurological damage.

In *Principles*, James broaches the parallelist view explicitly in his discussion of Jackson and Meynert, blurring any differences between their concepts of representation:

> "All nervous centres," says Dr. Hughlings Jackson, "from the lowest to the very highest (the substrata of consciousness), are made up of nothing else than nervous arrangements, representing impressions and movements. . . . I do not see of what other materials the brain *can* be made." Meynert represents the matter similarly when he calls the cortex of the hemispheres the surface of projection for every muscle and every sensitive point of the body. The muscles and the sensitive points are *represented* each by a cortical point, and the brain is nothing but the sum of all these cortical points, to which, on the mental side, as many *ideas* correspond. *Ideas of sensation, ideas of motion*

are, on the other hand, *the elementary factors out of which the mind is built up by the associationists in psychology.* There is a complete parallelism between the two analyses.[38]

It is this parallelist construal of the reflex arc that Dewey was to capture with his idea that it was "neither physical (or physiological) nor psychological" but rather "a mixed materialistic-spiritualistic assumption."[39] But as seen, there were in fact important differences between Jackson and Meynert on this very issue. Jackson denied that it was "muscles and sensitive points" that were represented at corresponding "cortical points"; rather what were represented were functions.

Despite James's general affirmation that "there is no doubt that [this parallelist view] is a most convenient, and has been a most useful, hypothesis, formulating the facts in an extremely natural way," it was not a view to which he held, at least in its standard form: "We shall have later to criticise this analysis so far as it relates to the mind."[40] And while James does not distinguish between the two neurological authorities here, it is clear that his misgivings about parallelism are tied to his misgivings about Meynertian associationism with which it was by then commonly collapsed. In *Principles* the sources of James's resistance to the parallelist solution to the mind–body relation is connected with two linked aspects of his overall stance: first, his support of the cerebral reflex idea with its postulated continuity between the cortical processes with the reflex action of lower centers, and next, his well-known thesis concerning consciousness as a continuous and changing stream.

The cerebral reflex thesis, as we have seen, on the one hand made somatic reflexes more "intelligent" and cerebral processes more mechanical than the Meynertian model in which somatic reflexes are fixed and served as mechanical implementers of voluntary actions originating in the cortex. James was impressed by empirical evidence of the spontaneous and functional actions of the somatic reflexes in lower animals—those characteristics of somatic reflexes that had made some researchers, including Lewes, entertain the notion of a "spinal soul." In *Principles* for James, these "mechanisms" were such that "the expression of some of them makes us doubt whether they may not be guided by an intelligence of low degree" (I. 75). James even speculates whether there may be a type of "split off consciousness" of the lower centers distinct from and unknown to "cortical consciousness" (I. 66–67).

Such considerations, he thought, posed problems for the Meynertian scheme for the lower animals, but these were minor compared to those facing its application to higher animals such as monkeys and humans. Here we find that "the hemispheres do not simply repeat voluntarily actions which the lower centres perform as machines. There are many functions which the lower centres cannot by themselves perform at all" (I. 75). There was in this respect a striking difference between lower and higher animals concerning the consequences of damage to the cortex. Thus, while "hemisphereless frogs moved spontaneously, ate flies, buried themselves in the ground," and so on (I. 73), "when the motor cortex is injured in a man or a monkey genuine paralysis ensues, which in man is incurable, and almost or quite equally so in the ape. . . . Even in birds and dogs the power of *eating properly* is lost when the frontal lobes are cut off" (I. 75–76).

Evolution seems to have been accompanied by a "passage of functions forward to the ever-enlarging hemispheres" with the effect that the lower centers in higher animals have become progressively more mindless and machinelike. For its part, the expanded cortex of higher animals is the inheritor of those more intelligent lower reflexes in the lower orders from which they have evolved: "The reflexes, on this view, upon which the education of our human hemispheres depends, would not be due to the basal ganglia alone. They would be tendencies in the hemispheres themselves, modifiable by education, unlike the reflexes of the medulla oblongata, pons, optic lobes and spinal cord" (I. 79–80). That is, the hemispheres are not Meynert's "virgin organs," which are "unorganized at birth." Rather, "they must have native tendencies to reaction of a determinate sort . . . tendencies which we know as *emotions* and *instincts*" (I. 76).

Such a picture of hard-wired connections in the cortex, native rather than acquired, now undermines the parallelism between neural and psychological structures as Meynert's "virgin" cortex was the psychical equivalent of a psychological *tabula rasa*. But on phenomenological grounds James was independently unhappy with the type of Lockean associationism assumed by Meynert. While the associationism of Meynert demands "ideas of sensation, ideas of motion" and so on as the "elementary factors out of which the mind is built up," factors that form a series able to parallel facts about the brain, James famously asserted that: "Consciousness . . . does not appear to itself chopped up in bits. Such words as 'chain' or 'train' do not describe it fitly as it presents itself in the first instance. It is nothing jointed; it flows. A 'river' or a 'stream' are the metaphors by which it is most naturally described" (I. 239).

The supposition held by almost all earlier analysts (James mentions especially the influence of Locke among the British and Herbart among the Germans) that mental life is constructed from discrete atoms of sensation, is condemned as amounting to an abandonment of the empirical method: "No one ever had a simple sensation by itself" (I. 224). The atomistic assumption had wreaked havoc in psychology.

It is precisely this assumption that the contents of consciousness can be broken down into quantifiable component parts that James objects to in Fechner's idea of a mathematizable psychophysical lawlike relation between the physical stimulus and the subjective sensation:

> Before the connection of thought and brain can be explained, it must at least be *stated* in an elementary form; and there are great difficulties about so stating it. To state it in elementary form one must reduce it to its lowest terms and know which mental fact and which cerebral fact are, so to speak, in immediate juxtaposition. . . . Between the mental and the physical minima thus found there will be an immediate relation, the expression of which, if we had it, would be the elementary psycho-physic law.
>
> Our own formula escapes the unintelligibility of psychic atoms by *taking the entire thought . . . as the minimum with which it deals on the mental side.* (I. 177)

Thus it is the putative combinatorial atomism presupposed by Fechner's parallelism that James seemed to oppose. One senses that if the parallelist doctrine could be liberated from the combinatorial framework and based on the minima of "the entire thought," then it would correctly capture the relation of the mind to the body.[41]

An important point concerning James's early provisional acceptance of the parallelist view should be noted here. James's criticism of the account of experience put forward in traditional empiricism is grounded in an objection that is itself a classically empiricist one. Earlier empiricists had, for James, simply not consulted their experience in a sufficiently careful manner. If they had, in their introspection, attended to the nature of inner sense, they would simply "see" that it formed a continuum. But it is precisely this overriding introspectionist assumption that James was to move away from in his later work, and there we find a new, quite different conception of a relationality within the realm of experience replacing the earlier idea of the introspectible "stream." This will modify his relation to classical nineteenth-century parallelist conceptions of the mind–brain relation even further.

James's Later Critique of Phenomenalism

It would seem that rather than having no place for cognition in his theory of emotion, James simply presupposed a different conception of the nature of cognitivity and its relation to somatic states from that allowed by the "platonizers." But exactly *what* this conception is is difficult to discern. Nevertheless, even at the time of writing *The Principles of Psychology,* an even deeper strand of James's anti-Cartesian approach was emerging in his writing, a strand that became more pronounced in his essays on "radical empiricism" after 1900.

In a paper, "The Function of Cognition,"[42] published just one year after the publication of "What Is an Emotion?" James further broached the question of the phenomenal dimension of cognition: "Cognition," he starts there, "is a function of consciousness. . . . Having elsewhere used the word 'feeling' to designate generically all states of consciousness considered subjectively, or without respect to their possible function, I shall then say that, whatever elements an act of cognition may imply besides, it at least implies the existence of a *feeling*" (179). Here we see James reasserting his anti-Platonist conception of cognition as rooted in the sensuous order, "feeling," and sounding a Nagelian note concerning the essential subjective locus of cognition—all cognitive functions of consciousness have a subjective "feeling" to them. Moreover, it becomes apparent that here his discussion of the cognitive functions of consciousness is from the "intentional stance": "cognitive feelings" in contrast to "simple" ones are "self-transcendent," that is, are intentional or semantic, pointing to or aiming at things other than themselves in the way that a gun aims at a target.

But what is intentionality, and what is it that allows certain mental states or "feelings" to hit their targets? First, there can be nothing intrinsic to the feeling that would guarantee its self-transcendence or reference: "The [self-transcendent] function is accidental; synthetic, not analytic; and falls outside and not inside its being" (186). Thus, James rejects the Lockean idea that *resemblance* between the feeling and that which it "knows" or refers to can provide the relevant link: "Eggs resemble each other, but do not on that account represent, stand for, or know each other" (187).

The way forward is indicated by virtue of an analogy. Consider a dream in which a certain event dreamt was later found to have actually occurred. "How," he asks, "would our practical instinct spontaneously decide whether this were a case of cognition of the reality, or only a sort of marvellous coincidence of a resembling reality with my dream?" For James, our practical selves would consider the dream a case of cognition if certain conditions applied: if the dream version agreed "point for point" with every feature of the real event; if, rather than being a single occurrence, it were one of a series of dreams all of which so coincided with real events; and, finally and crucially, if it allowed the dreamer "the power of *interfering* with the course of the reality, and making the events in it turn this way or that" (188). This is how one should think of the question of the self-transcendence of one's mental states, one's "feelings." Like the dream, we look to the feeling's consequences for the real world. "All feeling is for the sake of action, all feeling results in action—to-day no argument is needed to prove these truths" (189).

For James such considerations imply that the question of the self-transcendence of feelings is to be settled intersubjectively. Some other person's external point of view on the cognitivity of my "feeling" is needed because the self-transcendent function of the feeling "falls outside and not inside its being." (From within my dream I might, of course, dream that I could intervene in reality on the basis of my dreams!) That is, verification of whether or not a feeling hits its target can be obtained only from some reflective position (the position of what James calls "the critic") for whom the first subject is an embodied and acting worldly presence whose actions can be assessed in terms of whether or not they hit their targets: "If your feeling bear no fruits in my world, I call it utterly detached from my world; I call it a solipsism, and call its world a dream-world" (189). But surely, we might think, an opponent could raise objections here. First, how can "the critic" have "access" to the first subject's subjective feelings, so as to be able to identify that which is to be judged in terms of its success in hitting its target? And next, how does the critic him- or herself know that his or her own "feelings," that is, the mental contents with which he or she knows the targets of the first, are in fact themselves cognitive? (How can the critics be assured that their world is not a dream?)

James's implicit answers here reveal the distinctive and nonconventional conception of the mind and its intentionality presupposed. To pose the problems in such a way is simply to put things the wrong way around. "As a matter of fact, whenever we constitute ourselves into psychological critics, it is not by dint of discovering which reality a feeling 'resembles' that we find out which reality it means. We become first aware of which one it means, and then we suppose that to be the one it resembles. We see each other looking at the same objects, pointing to them and turning them over in various ways, and thereupon we hope and trust that all of our several feelings resemble the reality and each other" (190).

The general picture sketched here seems to be something like the following. We should not start from the conception of conscious subjects as immediately acquainted with their own private mental contents, their own "feelings," about which they could then ask the question of their cognitivity or self-transcendence. If one were to start here, there would be no way that one could establish the cognitivity of those states because the self-transcendent function is "synthetic" and "falls outside and not inside [the feelings] being." But such a solipsistic conclusion should be taken more as a *reductio ad absurdum* of the starting point. The charge of solipsism could practically be brought by a critic against some other person only on the basis that their would-be cognitive feelings "bore no fruit" in the shared world. It is from this shared world that we must then start. Rather than existing cognitively inside a world of our own feelings, our initial self-conception should be that of ourselves as already existing knowingly in a world of things, things we unproblematically share cognitively with others. "In the last analysis, then, we believe that we all know and think about and talk about the same world, because *we believe our* PER-CEPTS *are possessed by us in common"*(195–96). This was to be the core of James's later "direct realist" view of perception.

In these articles James was trying to approach the mind in a way that would circumvent the well-worn oppositions between subjective and objective, cognitive and physical, or first-person and third-person accounts. This attempt would result in the "radical empiricism" of later essays such as "Does 'Consciousness' Exist?," in which James, much like Daniel Dennett later, would reject the idea of the mind's "internal theatre," and in fact undercut the phenomenalist basis of his own earlier criticism of a combinatorial account of consciousness.[43] Yet, in James this approach would not be purchased at the expense of a sense of the reality of the "phenomenal consciousness" of mental states because James presupposed a vastly different conception of the phenomenal nature of experience to that which was traditionally presupposed. The most explicit accounts of radical empiricism are to be found in works twenty years after James's essay on the emotions. Nevertheless, as Gerald Myers has stressed, "the idea [of the later radical anti-Cartesianism] had been attractive to him even some years prior to the completion of *Principles*" (that is, *The Principles of Psychology* published in 1890). Moreover, Myers names "What Is an Emotion?" as one of "two essays . . . directing James towards the metaphysics of radical empiricism."[44]

In his most polemical denial of the existence of consciousness in the later "Does'Consciousness' Exist?," James makes it explicit that what is being denied is consciousness considered as an *entity* (one could equally say "realm"); but he is insistent on the reality of consciousness considered as a *function*. Central here is his denial of the sort of "mind-stuff" out of which mental contents (ideas, representations, and so on) could be made. Such mental entities could be abandoned because James had come to adopt a nonrepresentationist notion of conscious experience and its intentionality. In short, the mind's basic experiential relation to the world is not one mediated via the modification of mind-stuff—perceptual "representations." Rather, it is more correct to say that for James, as for later "ecological" theorists of perception such as J. J. Gibson, worldly things are directly and primarily *presented to* the mind.[45]

Such a presentationalist view of intentionality had been neatly presented in a short article, "The Tigers in India," published ten years after

"The Function of Cognition."[46] When we think of tigers in India, James asks, what is it about our state that makes it *tigers* that our thoughts are directed to? One answer might be that it was something about our mental representations themselves, in modern terminology, their semantic properties; but James repeats what he had earlier declared about "feelings," that is, that "there is no self-transcendency in our mental images *taken by themselves*" (200). Rather "The pointing of our thought to the tigers is known simply and solely as a procession of mental associates and motor consequences that follow on the thought, and that would lead harmoniously, if followed out, into some ideal or real context, or even into the immediate presence, of the tigers" (200). We might think of James as applying to the meaning of mental representation the dictum that Wittgenstein applied to linguistic ones, the injunction to look to their *use* in actual forms of life. From James's pragmatist perspective, the value or significance of our mental representations, of tigers, say, lies in the roles they play in our lives; having thoughts of tigers, for example, enables us to act in such ways that we might take "a voyage to India for the purpose of tiger-hunting" (200).

But if the value of representations lies in their ability to mediate forms of behavior that bring us into the presence of the things themselves, how then are we to conceive of our direct perceptual experience *of* such immediate presences? Surely, some nonrepresentational conception of direct perceptual experience must be presupposed here, and this is the inference that James is willing to make. In James's other example, in my direct experience of this white sheet of paper before me, it is this paper itself, and no mental simulacrum or representation of it, that is in my mind: "The paper is in the mind and the mind is around the paper, because paper and mind are only two names that are given later to the one experience, when, taken in a larger world of which it forms a part, its connexions are traced in different directions. *To know immediately, then, or intuitively, is for mental content and object to be identical.* This is a very different definition from that which we gave of representative knowledge."[47]

How are we to understand this very peculiar idea of the object, rather than its ideational representative or surrogate, as being in the mind? We might, of course, construe it as a striking *reductio* of the Cartesian assumption of the mind as having its own peculiar "extension" within which anything can exist, the image of the mind as a container being challenged by that of an extensionless point. But even though this may be a useful device for making us aware of a picture that has "held us captive," it hardly can count as an intelligible answer to the question of how to think of the mind's intentional capacities. Slightly more positively we might see it pointing in the direction of concepts used earlier by idealists

(Kant's use of the phrase of "consciousness as such" or Hegel's notion of a communicatively borne social *Geist*, or spirit) or later by "externalist" theorists of mental content for whom cognitive states cannot be individuated without reference to those things and events in the world that those states are about. But as fertile as James's suggestive formulae may be, they do not seem to be more than that—banners under which some radically non-Cartesian view of the mind might be pursued, rather than well-grounded conceptions.

Another way of taking James here might be to see him as making a similar point as had earlier been made by Reid in strictly differentiating the order of sensation from that of perception. Thus in "The Tigers in India," James suggests a conception of the relation of subjectivity to objectivity by conceiving them as two "great associate systems" through experience. A diagram shows a single horizontal line intersecting three vertical lines meant to stand for the mental history of three different persons. The points at which the horizontal intersects the three verticals are indicated with an "O" meant to represent a single object that can be conceived in common by the three persons. Considered as a point on any of the vertical lines, "O" is a part of a private mental history, but regarded as a point on the horizontal line, it "ceases to be the private property of one experience, and becomes, so to speak, a shared or public thing."[48]

The idea is repeated in "Does 'Consciousness' Exist?," where the object of the earlier paper is now the room in which the reader sits. On the one hand the room is "what it seems to be, namely, a collection of physical things cut out from an environing world of other physical things," but on the other it is "just *those self-same things* which his mind, as we say, perceives." But "the puzzle of how the one identical room can be in two places is at bottom just the puzzle of how one identical point can be on two lines." As in the earlier paper one of these lines "is the reader's personal biography, the other is the history of the house of which the room is part." From the former perspective that which is presently experienced "is the last term in a train of sensations, emotions, decisions, movements, classifications, expectations, etc., ending in the present, and the first term of a series of similar 'inner' operations extending into the future, on the reader's part."[49]

In the earlier paper James had just spoken of the two "associate systems," but now characteristic differences between these two systems are described:

The physical and the mental operations form curiously incompatible groups. As a room, the experience has occupied that spot and had that environment for thirty years. As your field of consciousness it may never have existed until now. As a room, attention will go on to discover endless new details in it. As your mental state merely, few new ones will emerge under attention's eye. As a room, it will take an earthquake, or a gang of men, and in any case a certain amount of time, to destroy it. As your subjective state, the closing of your eyes, or any instantaneous play of your fancy will suffice. In the real world, fire will consume it. In your mind, you can let fire play over it without effect. As an outer object, you must pay so much a month to inhabit it. As an inner content, you may occupy it for any length of time rent-free. If, in short, you follow it in the mental direction, taking it along with events of personal biography solely, all sorts of things are true of it which are false, and false of it which are true if you treat it as a real thing experienced, follow it in the physical direction, and relate it to associates in the outer world. (14–15)

With James's late philosophy of pure experience his critique of traditional empiricism has become more extreme, as it has severed its connection to the introspectionist framework that it had hitherto shared with that empiricism. But we still see the focus on the intrinsic connectedness or relationality of mental contents, that connectedness that had been linked to the early notion of the educable cerebral reflex, transposed to the new framework. A particular mental content, say the idea of tigers in India, has the particular identity that it has, that is, is a representation of tigers in India, because it belongs to a system of "substitutions" connecting that mental content with the content of certain perceptions—tigers in a particular shared location, India. What seen from one side is "a train of sensations, emotions, decisions, movements, classifications, expectations, etc., ending in the present" is, regarded from the other side a path traveling through the environment. One might well invoke the analogy of Fechner's famous curved line: seen from one side we are in contact with mental items, from the other actual worldly objects, but the model is now applied very differently. For Fechner, what, when seen from one side was a sensation, was, from the other, seen as a state of stimulation of the nervous system. But for James, what, when seen from one side is a sensation, is from the other, the worldly object that that "sensation" is usually thought to be a representation of.

Thus we can see that he now has a very different basis for an objection to the traditional parallelist picture. In Fechner's conception there are two points of view that run in parallel because the transitions within each are bound to each other in some definite mathematical relation. This linkage assumes, of course, that each side has a definite value. But James has now taken the idea of parallel perspectives into the subjective side of Fechner's dual aspects, making the idea of the mental content as sensation with a definite value as just a way of seeing what is otherwise a worldly object, the thing the sensations is, in fact, about. Fechner's picture of the mind–body parallelism has now been cut across orthogonally, as it were, by a quite different parallelism between subjective thought or experience and the objects that such thought or experience is about. Fechner's mind–body parallelism, which had been based on the "phenomenality" of mental states, has been disrupted by the presence of another form of parallelism relating to the mind's intentionality or cognitivity—its capacity to know the world or have an objective content. We might now ask what James's conception of emotion might look like from within this later, more mature expression of his understanding of the nature of mental life.

"The Place of Affectional Facts in a World of Pure Experience"

In an essay published in May 1905, "The Place of Affectional Facts in a World of Pure Experience," James returns to the topic of emotions, significantly now linked to what he calls "appreciative perceptions," and he does so within the context of his conception of "pure experience." Significantly, his opposition to the idea that "anger, love and fear are affections purely of the mind" takes the same path as in the essay of 1884. "That, to a great extent at any rate, they are simultaneously affections of the body is proved by the whole literature of the James-Lange theory of emotion."[50]

The issue of appreciative perceptions to which that of the nature of emotions is bound itself concerns the traditional question of the ontology of evaluative qualities such as aesthetic ones, precisely those qualities which have typically received a subjective or "projectivist" analysis in modern culture. It is against this latter tendency that James will invoke his idea of the one realm of "pure experience." Just as the white sheet of paper in "The Tigers in India" could be considered as both mental item or objective reality, so too can the beauty of the beautiful object. Expanding on Santayana's notion of beauty as "pleasure objectified," he notes that: "The various pleasures we receive from an object may count as 'feelings' when we take them singly, but when they combine in a total richness, we call the result the 'beauty' of the object, and treat it as an outer attribute which our mind perceives. We discover beauty just as we discover the physical properties of things. Training is needed to make us expert in either line."[51]

The proponent of a subjectivist or projectivist approach to qualities like beauty will, of course, protest that beauty is not like other "objective" qualities. The paper of the earlier example can be put in "objective" relations with other things of the outer world—using James's own example, "whatever is hard interferes with the space its neighbours occupy." Is it not the case that beauty lacks any analog to such "hardness"? But here James invokes two different senses of the inner–outer distinction. If we regard the physical world as that which "lies beyond the surface of our bodies," it is true that a thing's aesthetic qualities will have no "objective" effects on its neighbors. The disgusting character of a mass of carrion "fails to operate" on the sun that still caresses it or on the breeze that "woos it as if it were a bed of roses." But it " 'turns our stomach' by what seems a direct operation—it *does* function physically, therefore, in that limited part of physics."[52]

But has this not anaesthetized the quality "beauty" by reducing it to what is merely a dispositional capacity to induce certain effects in beings belonging to "that limited part of physics"? To answer "yes" here would be to consider my body as "merely" outer. But "our body itself is the palmary instance of the ambiguous. Sometimes I treat my body purely as a part of outer nature. Sometimes, again, I think of it as 'mine,' I sort it with the 'me,' and then certain local changes and determinations in it pass for spiritual happenings. Its breathing is my 'thinking,' its sensorial adjustments are my 'attention,' its kinesthetic alterations are my 'efforts,' its visceral perturbations are my 'emotions'."[53]

We might now grasp something of the unfolding of James's original identification of emotion with the physiological states of the body once it has found a place within a conceptual context more adequate to his original intuition. The disgust I feel, say in the presence of carrion, is a state of the body, a state caused by the carrion and corresponding to the "turned stomach" of popular understanding. James's critics would argue that rather than it being the carrion that turns the stomach, it is something fundamentally cognitive, the subject's "idea," "knowledge," or "interpretation" of the carrion, or, to bring James's more recent critics into the picture, the belief that the carrion is revolting. But James's answer will be precisely to posit the "turning of the stomach" as an integral part of what it is to know that the carrion is revolting, or, perhaps more exactly, as an integral part of what it is to know the carrion *as* revolting. Platonists cannot see that anything as bodily as a turned stomach could possibly be part of me as knower: they see the body as something that can only be part of "outer nature," never identifying their own body as part of their epistemic "me." Of course one can relate to one's own body in this way, as something belonging to the side of the "known" rather than the "knowing" as it were, and this is how we grasp ourselves physiologically. But the body is ambiguous, and sometimes I sort "my" body on the side of the knowing "me," as something I am, as it were, rather than as something "I" happen to "have." As outer object its "visceral perturbations" are, when it is taken as part of the me, emotions. And such emotions can themselves be regarded as cognitive states. Such perturbations, like other "local changes and determinations" can pass for "spiritual happenings," that is, for forms of knowledge.

But it also seems to be implied that I can equally consider my own body, hitherto the means for the knowledge, as a knowable object in the world like the carrion. Adopting this perspective on my body, putting it in these sorts of relations, I can understand how the carrion has certain qualities (no longer cohering as "disgusting") that can exert particular effects on certain of its neighbors, such as my body. As a concrete embodied knower it might be said that I do have an "inside," those phenomenally rich feelings that I construe "representationally" as disclosing qualities of the object: the churning stomach that I take as a knowing of the disgustingness of the carrion. But this inside is not that of a Cartesian mind as much as an inside defined by the skin. Moreover, I can also adopt a more general, impersonal perspective onto these feelings and sensations and see them as "knowing" not something outside me, the qualities of the carrion, but my own somatic states and processes and their causal triggers. Here my bodily states are no longer states I know with, but states I know of.

The modern cognitivist direction taken in both philosophical and psychological thought about the nature of emotion has too readily dismissed James. Understood in the context of his evolving thought about the nature of mind James's theory of emotion cannot be condemned as ignoring the cognitivity of emotional states. Rather, it must be seen against the background of an attempt to bridge the somatic–cognitive divide that still marks modern cognitive thought itself. But this is not to say that James's theory can serve as an unproblematic alternative to cognitivism; at best, it surely can be understood as fascinatingly suggestive, as gesturing toward rather than achieving any clear articulation of these difficult issues. But it may be that James can lead us forward here only by first leading us back into a maze of issues that preoccupied much nineteenth-century thought about the mind.

Perhaps we can see the distinction between James's two approaches to the body and the self as a not-so-distant descendant of Kant's transcendental–empirical distinction. As "my" body, my body serves as a type of "transcendental" condition of my sensory knowledge of the world—that it can reveal the world to me is assumed. But as a knowable object in the world, this normativity drops out. The same body is involved, but oriented in two radically different ways within "pure experience."[54] But what James seems to lack here is the sense of Kant's question into the conditions under which a mental state, a sensation, say, can be taken as a "knowing" of its object. For James the orders of perception and sensation are two different "associate systems," but he says little about the nature of their respective organizations. Emotions, as we have seen, can disclose certain sorts of facts, the disgustingness of the carrion, for example, but the logic of such judgments will presumably differ from that of judgments about properties that disturb a "part of physics" not restricted to that beneath the judge's skin. If emotions belong to an intelligible order and participate in certain kinds of judgments, they might be considered as having some kind of "logic," but presumably not one based on the logic of belief, which takes more conventional, non "appreciative" judgments as paradigmatic.

From James we get little idea of the nature of the relevant units and their connections here, beside the traditional idea of "association" which would seem inappropriate given its dependence on the traditional idea of some sorts of mental "things" to be associated. And this theoretical lack in James's picture seems to parallel a similar lack of interest in the psychodynamics of his own life. Gerald Myers points out that despite James's elaborately developed descriptive awareness of the nature of his own, often disturbed, affective states, and despite his constant searching out of various types of psychological therapies, James was strangely indifferent to the question of their psychological causation.[55] But this, of course, was the great focus of concern for another great philosophical psychologist who was developing his distinctive theories around the same time, Sigmund Freud.

3

Freud, Affect, and the Logic
of the Unconscious

We know Sigmund Freud as propounder of a number of controversial "discoveries"—the ubiquity of the sex drive, the hidden meaningfulness of dreams, jokes, and symptoms, the Oedipal drama of the family, and, most of all, the unconscious. This last concept offers an account of the sorts of relations and processes relevant to those elements on James's vertical line that represent the emotions. The unconscious was for Freud a "region" of the mind that operated not only without consciousness, but also with its own distinctive "logic"—the primary process. Moreover, the unconscious overlapped to a large extent with that aspect of the mind that Freud in his later topography called *das Es*, the "it" or "id," a cauldron of highly affectively charged instincts and drives.

It is over the notion of the unconscious that Freud and James are commonly contrasted: Freud driving a wedge between consciousness and the mind, and James to a large extent preserving the traditional identity of consciousness and mentality. But as I argue, Freud's theory not only developed out of the same matrix of psychophysical views as did James's, but also it leads us back into the same constellation of philosophical and psychological views of the mind as did James's—Kant and the transformations undergone by Kantianism by those trying to divest it of its otherworldly character.

Freud and the Physiological Background to Psychoanalysis

On the basis of his work on the neuroses and normal psychological phenomena like dreaming, Freud posited the existence of unconscious mental
states and processes. With this he meant, of course, something more than
the descriptive thesis of the existence of mental states of which the thinker
was not currently aware. For example, at any particular time we can be
said to have very many beliefs or memories of which we are not, at that
time, conscious. Without any further distinguishing features, such mental
contents Freud classed as "preconscious." What characterized the unconscious contents and processes he was interested in had to be understood
in dynamic and systematic or structural terms. For Freud, unconscious
contents were distinct from preconscious ones because they were subjected to certain dynamic processes that prevented them from reaching
consciousness. Using Block's terminology we would say that they are prevented from being "accessed": they are not "poised" for being used as a
premise in reasoning or for rational control of action or speech.[1] But this
does not mean that they are not expressed in reasoning, action, or speech;
it is just that they are expressed in odd ways. For Freud, the very oddness
of these ways attested to the fact that they were subject to distinctive types
of mental process belonging to "the unconscious" understood as a system,
the "topographical" unconscious. This systematic distinction concerned
structural characteristics of the thought processes to which such contents
were subject: while the contents of the system "conscious-preconscious"
were structured by the logical norms of objective thought that he labeled
the "secondary process," those of the structural unconscious were articulated in terms of the "primary process."[2]

As Frank Sulloway has pointed out in his influential *Freud, Biologist of the
Mind*, psychoanalysis had its immediate roots in the physiological thought
of the last decades of the nineteenth century.[3] In this respect it was similar to
Jamesian psychology. Freud himself seems to testify to this genealogy
when, in *The Interpretation of Dreams*, he invokes the universal role played
by the neural reflex in psychical life: "Reflex processes remain the model of
every psychical function."[4] And given Freud's background we should not
find this surprising. As a medical student in the 1870s, and then in the mid-
1880s as a young neuroscientist working in his "Laboratory for Cerebral
Anatomy" he had come under the influence of Theodor Meynert, the
anatomist and psychiatrist whose neuroanatomically based associationist
account of cerebral functioning had provided the starting point for James in
Principles of Psychology. However, also like James, Freud reacted against the
"Meynert scheme" and in an early neurological monograph *On Aphasia*,

published in 1891, criticized Meynert's model in a way that showed the particular influence of Hughlings Jackson.[5] The significance of these criticisms, however, goes beyond scientific issues and reveals much about Freud's underlying philosophical assumptions and the sorts of philosophical problems he would face in his later psychoanalytic theorizing.

In *On Aphasia* Freud, like James in *Principles*, published one year earlier, criticized the atomistically understood ideas of neural projection, representation, and association of Meynert and adopted Jackson's more functionalist conception of cerebral localization. On the basis of more recent anatomical findings Freud contests Meynert's view that there exist direct sensory tracks from receptors at the body's surface to the cortex. This correction to Meynert has the effect of making cerebral function, as James had said, more "mechanical": "Thus the theory of the dominant role of the cerebral cortex has been disproved. On the other hand, some processes previously regarded as sub-cortical, have now been allocated to the cortex." That all afferent and efferent pathways to and from the cortex must now pass through *intermediary* nuclei which subserve reflex arcs means that Meynert's idea of any point by point mapping of sensation and muscle innervation onto cortical areas "can no longer be maintained."[6]

> We can only presume that the fibre tracts, which reach the cerebral cortex after their passage through other grey masses, have maintained some relationship to the periphery of the body, but no longer reflect a topographically exact image of it. . . . If it were possible to follow in detail the rearrangement which takes place between the spinal projection and the cerebral cortex, one would probably find that the underlying principle is purely functional, and that the topographic relations are maintained only as long as they fit in with the claims of function. As there is no indication that this rearrangement is reversed in the cerebral cortex to produce a topographically complete projection, we may suppose that the representation of the body periphery in the higher parts of the brain, and also in the cortex, is no longer topographical but only functional.[7]

It is therefore not surprising that eight years later in *The Interpretation of Dreams* we find Freud positing the cerebral reflex as "the model of every psychical function."[8]

This appeal to a functional rather than topographical account of neural organization, and the associated abandonment of any ultimate distinction between lower "reflex" and higher cortical associative functioning that Freud takes from Jackson undercuts the classical associationist character of Meynert's schema. For Meynert, the fixed nature of these primary projections to and from the cortex meant that such cortical areas could localize or correspond to fixed psychological elements or meanings with identities independent of the associations into which they entered. With the functional rather than topographical construal of these processes, however, the separable identities of such postulated atomistic sensory and motor elements are lost. "Is it possible, then, to differentiate the part of "perception" from that of "association" in the concomitant physiological process?" Freud asks, and answers bluntly: "Obviously not. 'Perception' and 'association' are terms by which we describe different aspects of the same process. . . . We cannot have a perception without immediately associating it; however sharply we may separate the two concepts, in reality they belong to one single process which, starting from one point, spreads over the whole cortex."[9]

Again, as was the case with James, this rejection of the sensory atomism of classical associationism would in turn have repercussions on a metaphysical conception that Freud inherited from nineteenth-century medical science, that of the Fechner–Lewes doctrine of psychophysical parallelism. In *On Aphasia,* Freud embraces such a parallelist conception of the relation of mind and brain. "The relationship between the chain of physiological events in the nervous system and the mental processes is probably not one of cause and effect. . . . The psychic is . . . a process parallel to the physiological, 'a dependent concomitant'."[10] In writing this last phrase in English, Freud acknowledges the influence of Jackson from whom it is taken.

As we have seen, the "parallelist" idea can be traced back to Schelling's "nature philosophy" of the early years of the nineteenth century and, ultimately, to Spinoza. But Freud, while assimilating a generally parallelist framework, at the same time undercut the elementarist and classical associationist assumptions that had gone with it in Fechner's psychophysics in particular. Later, he would refer to the "insoluble difficulties" of the parallelist idea, more than likely because of the difficulty of resolving his idea of the unconscious with it.[11] Furthermore, he would, I suggest, work his way through some of this philosophical muddle by recovering conceptions of the mind from earlier philosophy, this time from Kant. These conceptions allowed him to reopen and explore a territory that had been earlier opened up by post-Kantians such as Schelling and Hegel, that of the unconscious.

It was not until the mid-1890s, after the short years of collaboration with Josef Breuer, that Freud started to develop his distinctive therapeutic approach to the neuroses via the interpretation of unconscious mental processes, "psychoanalysis." This approach he explained within an overarching theoretical "metapsychology" consisting of three linked perspectives: the structural, the dynamic, and the economic or energetic. The last of these bore most clearly the imprint of Fechnerian psychophysical thought and of his earlier neurology. One such Fechnerian element was a type of neurophysiological version of the law of conservation of energy that Fechner had called the principle of constancy.[12] Combined with the parallelist idea of the mind–body relation the constancy principle yielded a theory of pleasure and pain. For Freud, it was the accumulation of energy produced by sensory stimulation in the organism above an optimum level that was experienced as pain whereas its subsequent reflex discharge in motor action was experienced as pleasure.

In *The Interpretation of Dreams*, Freud used these psychophysical ideas to explain the operation of those processes he had come to believe lie at the heart of the unconscious, the so-called primary process. This form of psychic functioning that obeyed its own peculiar logic originated in the capacity of an infant to gain temporary satisfaction from the distress of an ungratified hunger drive via the hallucination of the breast, that is, the object that had earlier satisfied that drive.[13] Freud's idea was that in infancy, quanta of accumulated energy produced endogenously from the instinct of hunger coursed through the nervous system "seeking" the pleasurable discharge that was to be found in those actions, sucking at the breast and ingesting the mother's milk, which satisfied the need. This reflex, however, because of the cerebral associations of its pathways, was subject to "education" by being linked to other reflex actions. In *Principles* James had used Meynert's schema to explain how in an infant an association forged between the visual image of a flame and the pain of a burn would subsequently inhibit that very grasping reflex that initially predisposed it to reach for the bright flame.[14] Henceforth, the visual image of the flame would reflexly produce the withdrawal reflex action initially associated with the pain. In *The Interpretation of Dreams* Freud argues for a structurally analogous dynamic developing by association among the reflexes of infant behavior. The feeding of the infant on response to its distressful crying forges an association between the visual stimulus of the breast and the pleasurable satisfaction, such that the painful stimuli of the hunger drive alone could come to trigger a stored memory trace of the breast. The result of this was that the hallucination itself was now sufficient to effect an efferent discharge of the accumulated energy:

> An essential component of this experience of satisfaction is a particular perception (that of nourishment, in our example) the mnemic image of which remains associated thenceforward with the memory trace of the excitation produced by the need. As a result of the link that has thus been established, next time this need arises a psychical impulse will at once emerge which will seek to re-cathect the mnemic image of the perception and to re-evoke

the perception itself, that is to say, to re-establish the situation of the original satisfaction. An impulse of this kind is what we call a wish; the reappearance of the perception is the fulfilment of the wish; and the shortest path to the fulfilment of the wish is a path leading direct from the excitation produced by the need to a complete cathexis of the perception. Nothing prevents us from assuming that there was a primitive state of the psychical apparatus in which this path was actually traversed, that is, in which wishing ended in hallucinating.[15]

Of course in such a situation the actual satisfaction of the need still depends on the presence of the nurturing mother and on the actual feeding. But the original hallucinatory satisfaction was conceived as important in the development of the infant's subsequent behavior. With maturation the organism would become faced with the situation of no longer being able to rely on having its needs satisfied automatically by others; rather, a type of reflex response to the instinct that could produce real satisfaction now needed to be developed: the organism itself must act in the goal-directed way so as to bring about satisfaction. The requisite energy for this is still to come from the endogenous stimulus, so any premature discharge had to be postponed, and the nervous energy rerouted through patterns of thought and then action able to bring about the real satisfaction of the need by the "use of movement for purposes remembered in advance."[16]

In this way Freud's picture in *The Interpretation of Dreams,* drawing on his own earlier neurological speculations in *On Aphasia,* was in line with that of James: cortical associations allowed the earlier simpler reflexes to be inhibited and more complex patterns of response to develop. But two aspects of Freud's neurological account were to bequeath to psychoanalysis its characteristic shape. First, these mature patterns of thought, those of the secondary process, developed out of or were somehow grafted onto the earlier ones of the primary process, relying on them for their energy.[17] Next, utilizing Jackson's ideas about the nature of the speech center, Freud conceived of the higher cortical reflexes serving the secondary processes as somehow utilizing the structures subserving the capacity for propositional speech.

It was such ideas, especially that of the role of reflexes subserving speech functions in the functional representational organization of the cortex, that undermined the more classical associationism of the Meynertian model of cortical structure and function. Jackson's idea of the mediating role played by the speech center in cortical representation could give to the latter the type of propositional form that could not be accounted for by Meynert's simple associative processes alone. Moreover, with this very move Freud's picture became more faithful to the Darwinian principle espoused by Meynert concerning the "double origin" of the brain in evolution: "The whole organization of the brain seems to fall into two central apparatuses of which the cerebral cortex is the younger, while the older one is represented by the ganglia of the forebrain which have still maintained some of their phylogenetically old original functions."[18]

Meynert's evolutionary view, with its linking of emotion to the phylogenetically earlier parts of the brain, was one of a string of accounts of the brain as layered in this way, up to and including LeDoux's concept of the "emotional brain."[19] Jackson, too, on the basis of symptoms of aphasia, had argued that different parts of the brain had evolved at different times, although he approached the idea more in terms of the brain's lateralization, advancing the view that areas of the right hemisphere were more primitive than those of the left.[20] Moreover, Jackson had noted that in some forms of aphasic speech loss, although the patients had lost all voluntary speech they nevertheless retained the capacity for involuntary verbal responses, often in the forms of oaths or swear words. This difference, he described in terms of the difference between propositional and emotional speech, effectively appealing, as Anne Harrington has pointed out, to a conception that J. G. Herder had used a century before in his *Essay on the Origin of Language* (1772).[21]

The Nineteenth-Century Background to the Freudian Unconscious and Its Primary Process

The ideas concerning the distinction between the evolutionarily later "propositional" speech and the earlier emotional expression that Freud adopted from Jackson in the context of his neurological theorizing of the early 1890s were to underlie his later psychoanalytic distinction between the primary and secondary processes.[22] The outline of the story of the development of psychoanalysis over the last decade of the nineteenth century is well-known. Freud had been told by Breuer of a patient, "Anna O" (Bertha Pappenheim), who had seemingly been cured of numerous hysterical symptoms when, under hypnosis, she had remembered a traumatic event of her earlier life. Anna O herself had styled this phenomenon "the talking cure," and together Breuer and Freud developed "catharsis therapy" based on the hypnotic retrieval of memories. Freud, of course, eventually abandoned the hypnotism involved and developed a technique based on what was described as "free association," but like the idea of the separation of emotional and propositional speech, this too had been extant in late-nineteenth-century scientific and medical circles.

Freud played up the novelty of his idea of the unconscious mind and his technique of free association, but these were ideas with roots deep into the nineteenth century. In 1906 Carl G. Jung, the Swiss psychiatrist who at that stage was at the beginning of his short period as disciple and associate of Freud, traced the development of the type of word association test that was being employed at the Berghölzli clinic, where he was an assistant, and that seemed essentially related to Freud's own technique of free association.[23] Jung traced the technique back to a number of experiments that the English scientist and cousin of Darwin, Francis Galton, had performed on himself and subsequently published in *Brain*, the journal in which much of Jackson's work on aphasia appeared and to which Freud apparently subscribed. The technique had been quickly taken up and perfected by the German experimental psychologist Wilhelm Wundt, who attempted to establish laws of association of ideas. Subsequently, the psychiatrists Aschaffenburg and Kraepelin distinguished two types of associations, those that were made on the grounds of their meaning, and those based on purely "verbal" grounds, that is, similarities in sound alone. Aschaffenburg and Kraepelin noted shifts to the latter found in states of fatigue, fever, or alcoholic intoxication. A little later, the Berlin psychiatrist Theodor Ziehen measured the differentials in reaction time with the two types of associations and discussed these latter nonsemantic associations in terms of their belonging to "complexes" of representations that were "toned" or "colored" with feeling (*gefühlsbetonter Vorstellungskomplex*).

Such ideas were going to be merged with Freud's about the hallucinatory functioning of the distressed infant to provide a model of the affectively charged thought of the primary process. But even in the earliest appearance of these ideas in Galton's articles, these emotionally based associations had been described as for the most part working below the level of consciousness. These "trying and irksome" experiments, Galton reported, "gave me an interesting and unexpected view of the number of the operations of the mind, and of the obscure depths in which they took place, of which I had been little conscious before."[24]

Galton described his first experiment as involving a walk along Pall Mall, "during which time I scrutinized with attention every successive object that caught my eyes, and I allowed my attention to rest on it until one or two thoughts had arisen through direct association with that object; then I took very brief mental note of them, and passed on to the next object. I never allowed my mind to ramble. . . . It was impossible for me to recall in other than the vaguest way the numerous ideas that had passed through my mind; but of this, at least, I am sure that samples of my whole life had passed before me, that many bygone incidents, which I never suspected to have formed part of my stock of thoughts, had been glanced at as objects too familiar to awaken the attention."[25]

Galton repeated his walk and

> was struck just as much as before by the variety of the ideas that presented themselves, and the number of events to which they referred about which I had never consciously occupied myself of late years. But my admiration at the activity of the mind was seriously diminished by another observation which I then made, namely, that there had been a very great deal of repetition of thought. The actors in my mental stage were indeed very numerous, but by no means so numerous as I had imagined. They now seemed to be something like the actors in theatres where large processions are represented, who march off one side of the stage, and, going round by the back, come on again at the other.

To test this he cast around for a method that would allow some statistical analysis and created a series of words on sheets, devising an apparatus that could record the first associations that came to his mind when a single word presented itself. Galton reported that the figures showed "much less variety in the mental stock of ideas than I had expected, and makes us feel that the roadways of our minds are worn into very deep ruts. . . . I conclude from the proved number of faint and barely conscious thoughts, and from the proved iteration of them, that the mind is perpetually travelling over familiar ways without our memory retaining any impression of its excursions."[26]

With these "psychometric experiments," Galton was perhaps unwittingly reviving the earlier romantic interest in unconscious thought processes that, it was believed, were manifest in a variety of abnormal states of consciousness such as dreams, hypnoid and trance states, and various types of psychopathology.[27] A clear example of this earlier "romantic" medical tradition can be seen in G. H. Schubert's small book of 1814, *The Symbolism of Dreams*, a work to which Freud refers in *The Interpretation of Dreams*.[28] There Schubert, a former student of Schelling who had gone on to study and then practice medicine, talks of the soul expressing itself in dreams in a language that differs from that of waking life. For example, in dreams, things or properties of things can represent people, and in turn people can represent qualities or objects or actions. Furthermore, structurally, the language of dreams shows its own peculiar "laws of association," having a somewhat "hieroglyphic" nature in which pictorial elements are juxtaposed simultaneously or in quick succession. This gives to dream thoughts a more rapid and excited path or flight than that found in the thoughts of waking life, which finds its articulation in words.[29] In short, Schubert's "peculiar laws of association" resemble in many ways the later principles of the Freudian primary process, the logic-breaking laws of "condensation" and "displacement."

Although the early wave of "romantic medicine" had passed to be replaced by the more positivistic outlook characteristic of the "Helmholtz" school of medicine,[30] romantic notions were still extant in the general culture. These had been partly revived through the late reception of Schopenhauer and by late popularizers of Schellingian and Hegelian ideas such as Eduard von Hartmann, whose *Philosophy of the Unconscious* of 1869 had gone through nine editions in Germany by 1882.[31] These interests in abnormal states of consciousness blended into "spiritualistic" pursuits of seances, mediumism, automatic writing, and so on, which had earlier been justified on the basis of nature-philosophical ideas and which seemed to undergo a renaissance toward the end of the century. Indeed, a confluence of these ideas can be seen in the interests of individuals such as Frederick Myers, the founder of the "Society for Psychical Research," who knew Jackson through his brother Arthur, an epileptic patient of Jackson's who went on to become a physician and colleague of the neurologist. Jackson himself is said to have had a "great belief in the value of sitting quietly at the end of the day, notebook in hand, to allow unconscious reasoning to form new associations in his mind and to mull over the intellectual implications of his clinical experiences."[32]

Nor should we too readily divorce Jackson the scientist from earlier idealist and nature-philosophical traditions simply because of the availability of Darwinian materialist models for understanding the development of the nervous system. Jackson's views on evolution appear to have been derived more from Herbert Spencer than Darwin, and, through Spencer, from an Edinburgh- and London-based idealist tendency within medical and natural history circles in the 1830s, 1840s, and 1850s.[33] Particularly relevant here seems to have been William Carpenter, the unitarian physiologist and friend of Charles Darwin, who, as seen, along with Laycock had advocated the idea of the cerebral reflex. It was through Carpenter that Spencer had become influenced by the ideas of the major nineteenth-century embryologist, the Baltic German Karl Ernst von Baer, who described embryological development in terms of the differentiation and specialization of tissues.[34] Coupled with the idea of the reflex action of all neural processes, the idea of progressively functionally differentiated nervous tissue as one "ascended" the ganglionic strata of the central nervous system could give the hierarchical picture of neural functioning found in Jackson and Freud.[35]

James, too, was heir to this earlier romantic concept of the unconscious and was of course aware that certain forms of psychopathology were being discussed as involving "unconscious thoughts." But James had understood this idea in the way that had been put forward by Pierre Janet. To say that a person had unconscious thoughts was equivalent to saying that there was *another* consciousness embedded within their own conscious mind thinking those thoughts, an idea we have already seen in James's conjectures about the possible consciousnesses belonging to subcortical centers. In was in this way that patients with multiple personality disorders were thought to have other personalities somehow embedded within their "host" personality.[36] In fact, with Janet and James, "multiple personality," a disorder that attracted much psychiatric and popular attention at the end of the nineteenth century as it has at the end of the twentieth, was the form of psychopathology that neatly exemplified the structure of the unconscious and its relation to consciousness. It seemed to give direct expression to the idea of unconscious, but otherwise normally functioning, minds with their own personality structure existing or embedded, as it were, within the mind of the host.

Freud significantly departed from James, however, in his development of the idea of unconscious processes of the mind, and a central factor in their respective trains of thought was a notion that geared Freud's thinking into the distinctly Kantian issue of the role of representation and judgment in cognitive function.

Freud's Kantianism

In his version of empiricism, James, as seen, criticized the traditional philosophical "representative" theory of perception in which the immediate object of perceptual awareness was thought to be some distinctively *mental* item, such as a sensory "idea," which stood in a relation of "representation" to some external object or property in the world. For James, the notion of a psychic "representation" could apply only *within* experience: one mental content, my thought, say, of tigers in India, could only be said to represent some other experiential content such as an immediately perceived tiger encountered when my thought had played the appropriate role in leading me to that latter experience. A truly representational content was thus a "substitution" for a direct perceptual encounter, and its cognitive value was dependent on that of direct perception. This meant that actual perceptual encounter could not itself be thought of as a "representation" of a perceived object as in the traditional representative theory, but more as its direct *presentation*.

This antirepresentationist approach to experience provided an ambiguous legacy for James: on the one hand, it gave his theory a radically anti-Cartesian twist; on the other, however, James's conception of thought remained Cartesian in that it took consciousness as an essential characteristic of the mental. It was this latter aspect that led to his Janetian assumption that the unconscious should be thought of as a type of embedded "other" consciousness, a consciousness of which one was not conscious.

Freud's conception of the unconscious, however, stood in marked contrast to this and was based in his treatment of a different form of psychopathology, that of hysteria. In the context of their "cathartic therapy," Freud and Breuer reasoned that it was the cathartic discharging of accumulated affective energy associated with those memories that effected the cure because the undischarged energy itself was at the basis of symptom formation. Unable to be discharged in the normal way, the energy had sought an alternative expression in the symptoms, which constituted bizarre bodily forms of unconscious symbolic expression acting according to the "logic" of the primary process and that expressed a more primitive form of functioning of the brain. Against this background assumption, Freud could now raise the peculiarly Kantian question of the conditions needed for such functioning to be conscious.

"The Unconscious"

In his paper of 1915, "The Unconscious," Freud criticizes Janet's style of analysis of the unconscious as another consciousness and goes on to remark:

> In psycho-analysis there is no choice for us but to assert that mental processes are in themselves unconscious, and to liken the perception of them by means of consciousness to the perception of the external world by means of the sense-organs. We can even hope to gain fresh knowledge from the comparison. The psycho-analytic assumption of unconscious mental activity appears to us . . . as an extension of the corrections undertaken by Kant of our views on external perception. Just as Kant warned us not to overlook the fact that our perceptions are subjectively conditioned and must not be regarded as identical with what is perceived though unknowable, so psycho-analysis warns us not to equate perceptions by means of consciousness with the unconscious mental processes which are their object. Like the physical, the psychical is not necessarily in reality what it appears to us to be.[37]

Such a conception is radically removed from the idea found, say in Locke, Berkeley, or Hume, that the psychical must be "in reality" what it appears to be. Furthermore, it seems to contradict assumptions in Fechnerian psychophysical parallelism. In contrast to the usual view, Freud invokes the Kantian idea that a mental process involves "consciousness" only if it is itself the object of some further (presumably unconscious) mental act. Freud construes this further act on the analogy of an act of perception, a questionable interpretation of Kant's view.[38] Nevertheless, we might see emerging here the broad outlines of how a Kantian view of the mind can seriously bring into question the equation of mentality and consciousness, for if some act is required for some mental content to be conscious, that is, the act that ties that content into extensive relations with other contents, then two types of processes can, presumably, be unconscious, the act itself required for consciousness, and that original process when unaccompanied by that act.

For Kant, contents passively given to the mind are made conscious only in the context of the mind's own acts of synthesis, and he hints at the unconscious character of such acts, when, in the *Critique of Pure Reason* he describes the "imagination" as "a *blind* but indispensable function of the soul, without which we should have no knowledge whatsoever, but *of which we are scarcely ever conscious.*"[39] Following Kant, the idea that such synthetic acts are unconscious is even more explicit in Fichte and Schelling.[40] Here however, I want to focus on the latter issue, the idea of the *original* unconsciousness of some mental content or process when unaccompanied by some further unifying act. Essentially, such a view results in the idea of the unconsciousness of sensation, an idea that is at least hinted at in Kant, found in more explicit form in Schelling, and revived in recent philosophy of mind. It is this "Kantian" conception of consciousness with its unconscious substrata of sensation superimposed on Fechnerian psychophysical parallelism that, I suggest, results in the distinctively Freudian opposition between consciousness and the unconscious.

In the classical associationist conceptions of the mind sensations or "sensory ideas" are the paradigms of conscious mental content, being conceived as both states of the mind, and that which is known most directly by the mind, that is, what the mind encounters when it reflectively attends to its own contents. In the aphasia monograph, Freud had abandoned this with his criticism of the separability of sensory inputs from what had been already become associated with such inputs at the level of subcortical reflexes. But a century earlier Kant also had brought this view of sensation into question. While he thought that the turn to inner sense in introspection could indeed reveal a type of "natural description of the soul,"[41] it is crucial that for Kant the elements of such a descriptive psychology had to be thought of as intuitions already subject to the transcendental ordering of the pure intuition of time.[42] While sensations, in turn, provided the "matter" of thus formed "intuitions," sensations per se are not intuitable "objects" of inner sense. Sensation, Kant tells us in the *Critique of Pure Reason*, "is not in itself an objective representation, and since neither the intuition of space nor that of time is to be met with in it, its magnitude is not extensive but *intensive*." This quantitative intensity is what allows Kant even to suggest a Fechner-like proportional relation of sensation to stimulus: "Corresponding to this intensity of sensation, an *intensive magnitude*, that is, a degree of influence on the sense, must be ascribed to all objects of perception, in so far as the perception contains sensation.[43] In *System of Transcendental Idealism*, Schelling, in contrasting sensation and the objects of inner sense, would later refer to the latter as "sensation with consciousness,"[44] and the idea of the unconsciousness of sensation has returned recently in the work of a number of philosophers and psychologists working within a "higher-order thought" approach to consciousness.[45]

In fact, such an idea of unconscious sensations seems to take on an even more distinctly Freudian complexion when it is noted that Kant considers these sensations as themselves governed by physical laws of association in contrast to the conceptual or rule-governed ways in which intuitions are brought together in processes of "synthesis." Kant had criticized the "physiological" account of cognition found in Locke that tried to answer the question of the validity of judgments in terms of quasi-physiological postulates concerning how sensations became associated. Only a rule-governed synthesis of intuitions could produce judgments with an epistemic value. But this separation of association of sensation on the one hand and the rule-governed synthesis of intuitions on the other seems to have left Kant with a picture not at all unlike those of Kraepelin, Ziehen, and Freud. Indeed such a picture is reinforced by what Kant says in his *Anthropology from a Pragmatic Point of View* about the role of attention and abstraction in mental life.[46]

That we humans become conscious of our mental contents is, for Kant, not an event that automatically accompanies the mere having of mental contents; rather, it is an achievement, the outcome of the two mental activities of attending to (*attentio*) and abstracting from (*abstractio*) our mental contents, and these are cognitive activities that, especially in the case of abstraction, can be done with greater or lesser success.[47] Abstraction, he tells us, involves keeping a content "separate from other sense impressions in my consciousness. Therefore, one does not speak of abstracting (separating) something, but of abstracting from something, that is, abstracting a definition from the object of my sense impression, whereby the definition preserves the universality of a concept, and is thus taken into the understanding."[48] That the representation must be actively separated from others by this cognitive act presumably testifies to its tendency to become naturally associated with others on the basis of the "physiological" operation of the law of association. These processes, thought Galton and Freud, are actually going on all the time in the "antechamber of consciousness," irrespective of the contents of *conscious* thought. And something like this seems to be implied in Kant as well. For cognition to take place, the mind must somehow intervene within this layer of association and isolate a content from this stream via the acts of *attentio* and *abstractio*. It is only then that the content is able to enter into rule-governed syntheses that result in judgments. It is only then, too, that one is "conscious" in the full cognitive and "access" sense of the word.

The realm of this unconscious associative activity is that of the "lower" cognitive faculty, "sensibility," which passively receives its contents from the external world. In contrast, "intellectual cognition has the character of spontaneity of apperception, that is, of the pure consciousness of the act which constitutes thought and appertains to logic." But the spontaneous faculty, of course, needs the "immeasurable field" of the passive, receptive faculty to supply its acts with content, and so "on this account consciousness is divided into the discursive (which, being logical, must take the lead because it provides the rule) and intuitive consciousness." These two forms of "consciousness," which are thus both necessary for representations to become conscious, represent opposing poles of the self. "Discursive consciousness (the pure apperception of its mental activity) is simple. The 'I' of reflection contains no manifold within itself, and is always one and the same in every judgment, because it contains merely the formal part of consciousness. On the other hand, inner experience contains the material of itself as well as a manifold of the empirical, inner perception, that is, the 'I' of apprehension (hence an empirical apperception)."[49]

For Freud unconscious mental processes neither function within the framework of time nor obey logical laws, and the same must be true also for Kant; given their lack of "extensive" relations, unconscious sensations cannot be understood as existing within time or in connection to concepts. Schelling points to the problem of assuming sensation to be conscious when he notes that in sensation "the self is entirely rooted upon the sensed, and, as it were, lost therein."[50] That is, sensation could not function to represent or present some content, even itself, to or for the subject, for there exists no requisite gap for a content to "stand over" against that subject. Freud seems to imply such a progression from rudimentary sensation (akin to that of endogenous stimuli, for example) to genuinely representational mental states in his discussion of the development of the reality principle and the consequent heightened importance that is ascribed to the "sense-organs that are directed towards [the] external world" and to "the *consciousness* attached to them."[51]

Kant and Freud's Topography

The Kantian framework of Freud's thought is also clearly visible in his account of the "topographical" structure of the mind. After *The Interpretation of Dreams*, the differentiation of primary and secondary processes was further broached in a paper of 1911, "Formulations on the Two Principles of Mental Functioning." There Freud described the failure of the hallucinatory discharge of the primary process to satisfy actual needs as leading to its supplementation by the "reality principle," a principle that cognitively orients the organism to the state of the external world (and not just the internal states of its own pleasure) and that articulates the organism's motor discharges with that world in a goal-directed way. With the formation of this new principle of brain functioning the infant thus endeavors "to form a conception of the real circumstances in the external world and to endeavor to make a real alteration in them."[52]

For Freud this new type of functioning involves directing attention to qualities specifically given by the outer senses. "Consciousness now learned to comprehend sensory qualities in addition to the qualities of pleasure and unpleasure which hitherto had alone been of interest to it."[53] But to become relevant to the operations of a reality principle, such givens of external sense must, of course, be grasped cognitively; the organism's orientation to them has to be a matter of an "*impartial passing of judgement*, which had to decide whether a given idea [or representation, *Vorstellung*] was true or false," and this is achieved in terms of the demand for unity among these judgments, "the decision being determined by making a comparison with the memory-traces of reality." Thus, for this to be possible, the organism must have some means of laying down memory-traces, and so "a system of *notation* was introduced, whose task it was to lay down the results of this periodical activity of consciousness, a part of what we call *memory*."[54]

This appeal to the unification of judgments within experience as the measure of objectivity has a distinctly Kantian ring. For Kant, the cognitive question of the "objectivity" of the mind's representations turned on the question of their justification rather than, as Locke had thought, on that of their genesis. To focus on the question of justification means that the objectivity of appearances must be thought of as a matter *internal* to the realm of experience or appearance, not one concerning the relation of some experiential content to a "thing in itself" beyond experience. That is, the question of objectivity had to become translated into the question of the coherence or unity among the mind's actual judgments or representations, a unity tied to that of the mind itself, the "transcendental unity of apperception." While for Kant this latter unity of self-consciousness was conceived in a very abstract way—as simple, formal, "I" of reflection—many of those following him were concerned with the question of how a concrete individual subject was able to maintain the requisite awareness of its own particular unity. These included not only the well-known classical figures of idealism such as Fichte, Schelling, and Hegel, but also a thinker such as Johann Friedrich Herbart, a student of Fichte and Kant's successor at Königsburg, who was to become an important figure in the development of later psychological thought.[55]

Unconscious Emotion

It would seem that it was by posing such a Kantian question concerning the conditions of consciousness that Freud, more than James, could draw out the consequences of separating its cognitive and phenomenal dimensions and thereby develop his distinctive and multifaceted theory of the *unconscious*. In Freud we see the distinction between cognitive (or "access") consciousness and phenomenal consciousness emerge most clearly when he raises the question of whether or not emotions can be unconscious.

In "The Unconscious," Freud comments on the essentiality of the phenomenal character of feelings or emotions: "It is surely of the essence of an emotion that we should be aware of it, i.e., that it should become known to consciousness."[56] But, he notes, in psychoanalytic practice "we are accustomed to speak of unconscious love, hate, anger, etc., and find it impossible to avoid even the strange conjunction, 'unconscious consciousness of guilt,' or a paradoxical 'unconscious anxiety'." How can this be? How can there be an unconscious emotion if it is "of the essence of an emotion that we should be aware of it"?[57]

It is in Freud's answer to this question that we see the difference between phenomenal and cognitive or access aspects of consciousness explicitly appear. An affective or emotional impulse is necessarily felt, but may be misconstrued by being "attached to another idea" after the original idea to which it had been attached had undergone repression. Thus to talk of the presence of an unconscious emotion, say, unconscious guilt, is not to talk of the presence of that emotion without any "feeling" or phenomenal character; rather it is a matter of that phenomenal character not being understood as guilt, with all that that implies.[58]

Freud's idea seems to be the following. Say I feel guilty about some act, A. The *Vorstellung* "attached" to the phenomenal state will be some kind of representation, such as a memory of that act. This representation is repressed, that is, moved into that system of representation of the unconscious. But only representations are susceptible to this type of operation, and affects and emotions are not themselves "representations." Rather, they are more like the mental side of a single psychophysical reality: they "correspond to processes of discharge, the final manifestations of which are perceived as feelings." What then happens to them when their "attached" representation is repressed? Freud names three possible fates: the affect can remain totally or partially as such; it can undergo a transformation into qualitatively different "quota of affect," especially anxiety; or it can be suppressed and "prevented from developing."[59]

From this it seems clear then that to be "unconscious" a mental activity or process, specifically an endogenously originating emotional one, does not have to be *phenomenally* unconscious. It can be present in some way as "quale," as it were, but not present as articulated with those representational elements giving it cognitive value. Using Block's distinction, we might say that affects can be "phenomenally" conscious without being "access" conscious; using Kant's, we might say that they can be entirely intensive with no extensive relationships. With Freud's notion of the repression of the *Vorstellungen* associated with an emotion leaving the mute and undirected "feeling" of anxiety, we encounter a view of undirected or "free-floating" anxiety not unlike its construal as a "non-objectal" affective state.

Linking this to the previous suggestion concerning Freud's postulation of Kant-like conditions for consciousness, it might now be suggested that being the object of a "meta" apperceptive act is what is required for a phenomenally conscious mental process or content to become "conscious" in the full, cognitively rich sense of the word. It is this latter sense of conscious that enters into opposition with the notion of "unconscious." And the same "conscious" allows Freud to move to the idea that to describe mental contents and activities as "conscious" is to give them a location within a particular *system* within the mind, a system that can ensure the cognitive status of its contents and that structurally contrasts with another rival system, that of the unconscious. It is with this systematic aspect of consciousness, an aspect that in a further development of Freud's psychic topography becomes associated with the idea of the ego, that the Kantian dimension of Freud's thought becomes the most explicit. But what are the conditions that allow a merely intensive mental state to become conscious? In his answer, Freud here draws on ideas about language that Jackson had used in his studies of aphasia.

We have noted Freud's idea that the mind's acts are "essentially" unconscious and that to become conscious they must borrow something of that "consciousness" that accompanies the external senses. Freud, following Jackson, invokes the faculty of propositional speech as the key to this transmission of consciousness from that perceived externally to that "perceived" internally. Thus in *The Interpretation of Dreams*, "The Unconscious" and "The Ego and the Id" Freud discusses the importance of memories of "word-presentations" in the ego's functioning. Thought processes "i.e. those acts of cathexis which are comparatively remote from perception, are in themselves without quality and unconscious, and . . . attain their capacity to become conscious only through being linked with the residues of perceptions of *words*."[60] The ego's activity must be thought of as in some sense an internal act, but "only something which has once been a *Cs.* perception can become conscious, . . . anything arising from within (apart from feelings) that seeks to become conscious must try to transform itself into external perceptions: this becomes possible by means of memory-traces."[61] As such memory-traces can be derived only from earlier sensory impressions of perceived speech, the consciousness of inner sense is thus borrowed from that of outer sense.

Again, with this he returns to a tradition initiated by Kant, who in his *Anthropology from a Pragmatic Point of View,* had similarly described an individual's thought processes as inner speech: "All language is a signification of thought; the supreme way of indicating thought is by language, the greatest instrument for understanding ourselves and others. Thinking is speaking to ourselves. . . . Consequently, there is also hearing ourselves inwardly (by means of the reproductive imagination)."[62] What separates humans from other animals has to do with the aspiration to this higher, cognitive form of self-conception or "I-hood" by which experience can be unified. In line with the strong connection he makes between the capacity for conceptual thought and language, Kant emphasizes the significance of the child's ability to refer to itself with the word "I." It is by means of this act that the child goes from simply feeling itself to a self-conception, a passage from a self-relation that because it exists simply as feeling is entirely egocentric—an egoism—to one that because it is mediated by a concept allows one to understand oneself as one among other "I"s—pluralism.[63] Following Kant, such a linkage between the capacity for conceptual thought and language, was made even more explicit by various philological thinkers such as Wilhelm von Humboldt and Friedrich Schleiermacher and philosophers such as Fichte, Schelling, and Hegel.

Information and Energy in the Freudian Mind

Prior to the development of psychoanalysis Freud, following both Fechner and Jackson, had held to a psychophysical parallelist view of the mind–body relationship. With the development of the idea of the unconscious, however, such a form of parallelism had to be abandoned, seemingly because of the inability of psychophysical parallelism to provide a place for unconscious mentality. In the parallelist picture, all that could be "unconscious" is the workings of the brain, but Freud wanted the gaps in the conscious chain of psychological causes to be filled with something psychical but unconscious, not something physiological. But this is not quite right. Nineteenth-century physiological psychology could give some place to the idea of unconscious sensation as seen in J. F. Herbart's notion of the "limen."[64] Parallelists could understand this to mean that mental contents whose energetic states fell under a certain threshold would not become conscious. For the parallelist position, unconscious mental processes would be such because they were less energetic than conscious ones. But in Freud's picture the reverse was actually the case. Unconscious mental processes were typically more energetic than conscious ones.

As we have seen, in Kant not only is there the suggestion that unschematized "sensation" is access unconscious, but that its "intensity" corresponds to "an *intensive magnitude,* that is, a degree of influence on the sense [that] must be ascribed to all objects of perception."[65] It would seem that Kant, too, subscribed to a limited form of psychophysical parallelism, one that applied to that "immeasurable field" supplied by the passive, receptive faculty and from which cognitive contents are formed by processes of attention and abstraction. For Freud, too, as the discussion of the primary process in *The Interpretation of Dreams* attests, the unconscious works according to "energetic" principles. This aspect is a key factor that differentiates it from the cognitive processes of consciousness, which, it would seem, work according more to the principles of an "informatics."

In "Beyond the Pleasure Principle," Freud speculates about the evolutionary development of the sense organs, but reverses the assumptions from which such discussions typically start. First, he notes the fact that embryologically the cerebral cortex develops from ectodermal tissue, that is, that layer of cells on the embryonic plate that is in general the source of the outer, protective skin of the body. Thus, "the surface turned towards the external world will from its very situation be differentiated and will serve as an organ for receiving stimuli." Freud then construes the organism's original problem, as it were, as one of too much rather than too little stimulation. (Here we must remember Freud's Fechnerian assumption that stimuli are, for the organism, essentially *painful*.) Would it not be the case then that in the organism a "crust" or "shield" might have formed to protect it from the stimulation emanating from the powerful energies of its environment? "*Protection against* stimuli is an almost more important function for the living organism than *reception of* stimuli."[66] This conception then guides Freud in his thought of how exactly to think of the operations of the external senses:

> In highly developed organisms the receptive cortical layer of the former vesicle has long been withdrawn into the depths of the interior of the body [i.e., as the grey matter of the cerebral cortex], though portions of it have been left behind on the surface immediately beneath the general shield against stimuli. These are the sense organs, which consist essentially of apparatus for the reception of certain specific effects of stimulation, but which also include special arrangements for further protection against excessive amounts of stimulation and for excluding unsuitable kinds of stimuli. It is characteristic of them that they deal only with very small quantities of external stimulation and only take in *samples* of the external world. They may perhaps be compared with feelers which are all the time making tentative advances towards the external world and then drawing back from it.[67]

Freud then, with very little explanation, links this to the Kantian nature of the system he there calls *Pcpt.-Cs*, the system of perception-centered consciousness. In contrast to the "timelessness" of unconscious mental processes, conscious mental processes are within time. He then adds: "This mode of functioning may perhaps constitute another way of providing a shield against stimuli."[68]

Much the same line of thought is repeated in the later paper of 1925, "A Note upon the 'Mystic Writing Pad'." Perception should be thought of as a sequence of sensory samplings as the unconscious stretches out its feelers

> through the medium of the system *Pcpt.-Cs.*, towards the external world and hastily withdraws them as soon as they have sampled the excitations coming from it. . . .
>   So long as that system is cathected in this manner, it receives perceptions (which are accompanied by consciousness) and passes the excitation on to the unconscious mnemic systems; but as soon as the cathexis is withdrawn, consciousness is extinguished and the functioning of the system comes to a standstill.[69]

From his "sampling shield" model of external perception, Freud in "Beyond the Pleasure Principle" then contrasts external with internal perception, noting that

> the difference between the conditions governing the reception of excitations in the two cases have a decisive effect on the functioning of the system and of the whole mental apparatus. Towards the outside it is shielded against

stimuli, and the amounts of excitation impinging on it have only a reduced effect. Towards the inside there can be no such shield; the excitations in the deeper layers extend into the system directly and in undiminished amount, in so far as certain of their characteristics give rise to feelings in the pleasure-unpleasure series.[70]

Thus exogenous and endogenous sensation work in very different ways, and of the two, it is clear that Freud thinks that the internal, endogenous system alone works in a Fechner-like parallelist way. "The excitations coming from within are, however, in their intensity and in other, qualitative, respects—in their amplitude, perhaps—more commensurate with the system's method of working than the stimuli which stream in from the external world." In contrast, with their "sampling" operations, external senses working as they do with relatively low-energy states function in quite different, and apparently, non-Fechnerian ways. In contemporary cybernetic parlance we would say that for the external senses, effects are brought about not by the flow of energy, but by the flow of "information" that can be conveyed in the reduced flow of energy within the nervous system allowed by its "sampling" mode of operation. But this sampling-shielding mode of operation of the external senses can break down, as it does in traumatic experience: "We describe as 'traumatic' any excitations from outside which are powerful enough to break through the protective shield."[71]

We might expect that a system working on "informational" principles might somehow be able to become bound up with the operation of symbolic processes such as found in propositional language. Given the connection between consciousness, the ego, and language, it would seem that this is indeed the case.

The Structural Unconscious as Another Language of Thought

We have seen how the idea of the existence of unconscious aspects of mental life had been opened up as an area to be explored by Kant's philosophy. In fact, via various transformations, Kant's implicit concept of the unconscious had become explicit in the "nature-philosophical" speculations of Schelling. We have noted that Schelling's ideas had penetrated medico-psychiatric speculation about abnormal states of consciousness as can be seen in early works such as Schubert's book on dreams. In fact during the first two or three decades of the nineteenth century, nature-philosophical ideas proliferated throughout the emerging discipline of psychiatry in Germany. As one historian of the period notes: "At almost every German university, ideas on mental illness were expounded within systems of natural philosophy or religion, by physicians and/or philosophers."[72]

Many of these philosophers and physicians had, like Schubert, been influenced by Schelling, and Schellingian nature-philosophical ideas were spread throughout the German medical community by figures such as J. C. Reil, Karl Eschenmayer, Lorenz Oken, Ignaz Troxler, Henrik Steffens, J. C. A. Heinroth, C. G. Carus, and Franz von Baader. As we will see, Schelling's nature-philosophy evolved out of a line of thought grounded in the philosophy of Kant and Fichte as well as an intense interest in contemporary science, but as it both developed in Schelling himself and especially in those influenced by him it acquired progressively bizarre and irrational features.[73] An interest in altered states of consciousness such as found in dreams and hypnotism blended into spiritualistic and mystical interests. All phenomena, it was thought, could be captured in terms of Schelling's analogically projected "polarities," in which conscious and unconscious were seen as homologous with the alternation of day and night, life and death, male and female, Occident and Orient, and so on.[74] Nature philosophy had had its critics within the sciences from early on, such as the French comparative anatomist Georges Cuvier, who, in 1810 in his report to Napoleon's Council of State denounced it for its confusion of moral and physical realms and metaphorical and logical thought.[75] Later, the development of the "biophysics movement" in Germany in the 1840s and 1850s helped to kill it off by mid-century. Although nature-philosophical medicine is now commonly regarded by historians of science as having led to important breakthroughs, especially in areas such as neurophysiology, its reputation could not have been helped by incidents such as the publication of Justinus Kerner's *The Seeress of Prevorst: Revelations Regarding the Inner Life of Man and the Way a Spirit World Projects into Our Own* in 1829. In this immensely popular book, Kerner, by all accounts an effective and humane doctor as well as a poet, chronicled the progress and visions and prophesies of a patient, Friederike Haufe, who was thought to have access through her mental illness to the truths of the "nocturnal side" of life.[76] Great interest in the seeress was taken by Schelling, Schubert, Eschenmayer, and others, and her revelations evidently were taken seriously. After her death the interest in parapsychology was continued by Kerner and his circle with the journals *Blätter vom Prevorst* and *Magikon*.[77]

This spiritualistic movement underwent a revival in Germany toward the end of the nineteenth century, and it is now a common criticism of Freud to see him tainted by its irrationality. Indeed, as pointed out, there is more than a passing resemblance between the structures of the

Freudian primary process and the analogical and polar thought postulated by the nature philosophers as the language of the unconscious.
Moreover, not only was Freud aware of this tradition, but also in the period in which he was shaping his ideas concerning the nature of the primary process he was faced with the development of a similar approach to
these phenomena that was clearly working within a contemporary version of romantic psychiatry. This was the approach being developed by
Carl G. Jung. In fact, Freud's "Two Principles" appeared in 1911 in the
same edition of the *Yearbook for Psychoanalytic and Psychopathological Re-
searches*, the official journal of the first International Psychoanalytic Association, as a similar piece by Jung, "The Two Kinds of Thinking." Indeed,
Freud's article seems to have been written as a direct response to Jung's.[78]

The ideas of Galton, Wundt, Ziehen, and others formed the conventionally scientific context within which Jung developed ideas concerning the
two kinds of thinking. One centered on the semantic content of words,
whereas the other involved only the loose verbal associations of affect-
charged *Vorstellungskomplexen*. But with his particular interest in the typically "romantic" topics of mythology and the occult, Jung was taking such
ideas in a decidedly less conventional direction. "The Two Kinds of
Thinking" was in fact written as the first part of a longer work, the second
part of which appeared later under the title of "Transformations and Symbols of the Libido," and which would be mainly concerned with very
speculative ideas about mythological universals. The link was that such
affectively charged complexes of representations were common to
dreams, mythology, and the psychotic disorder of *dementia praecox* (Schiz-
ophrenia). Freud also, at this time, indulged in speculative psychoanalytic
readings of myths, but whereas he adopted a reductionistic attitude to
them, explaining them in terms of the pleasure principle and the operations of the "primary process," Jung, in some sense, seemed to be going
the other way, taking myths, dreams, and so on as symbolic expressions
that contained some deeper objective truth that could be decoded. He
even to some degree revived early-nineteenth-century nature-philosophical
ideas of superindividual minds by attributing a "phylogenetic" memory
to the species, that is, making memory itself something which went beyond the limits of an individual's experience.[79]

We can then see both Freud's and Jung's accounts of the unconscious as
providing models of what we might consider as a "language" or "logic"
of affect. However, one might appreciate Freud's concern over the price of
accepting the type of romantic model being developed by Jung, given its
connection with the wilder, speculative quasi-religious views of which
Jung has been accused. In contrast to this, Freud espoused the role of scientist and to some degree projected this attitude onto the process of thera-

peutic interpretation itself. The patient was to bring the ego and its reality principle to that psychic material that hitherto remained within the system of the unconscious. The slogan for this was: "Where Id [or "it," *Es*] was, there I [*Ich*] shall be."[80] In fact, we might take the opposition between Freud and Jung over the relations of unconscious and conscious thought as repeating an opposition a century before between Hegel and Schelling. But as it is Kant who stands at the start of this tradition, which allows talk of the unconscious, and as it is Kant who in the history of recent philosophy poses the question of the nature of mental unity in fundamental ways, we will turn now to a further examination of Kant's views of the nature of mind.

4

Kant, Mind, and Self-Consciousness

In the 1980s and 1990s a number of philosophers working within contemporary philosophy of mind and cognitive science rediscovered the importance of Immanuel Kant's thoughts about the nature of the mind.[1] Against such a background of the "mind's new science," it is not difficult to understand why this idealist writing almost two centuries ago could come to be regarded as having something important to say.[2] Kant might be seen as the beneficiary of a swing within early cognitivist thinking (and, especially within linguistics with Chomskian theory) away from the empiricism beloved of behaviorists toward more "rationalist" positions. He, too, had opposed empiricist and associationist approaches to the mind as found in Locke and Hume, replacing the idea of the mere compounding of ideas with that of the mind's rule-governed acts of synthesis.[3] But from the modern perspective Kant also looks much more up-to-date than the rationalists that preceded him. For example, he dealt with earlier conceptions of "faculty psychology" in essentially functional ways. As Ralf Meerbote has noted, Kant's notion of "transcendental" often seems to mean much the same as the modern cognitivists' use of "cognitive" or "epistemic";[4] and the strategy of "transcendental argument" also tends to appeal to the modern cognitivist's interest in the conditions necessary for the exercise of a capacity.[5] Furthermore, this once-derided faculty psychology now tends to be regarded differently in the light of the thesis of the "modular" character of the mind.[6]

Early suggestions of understanding Kant along the lines of the emerging functionalist or representationalist models of the mind can be found in the work of Wilfrid Sellars, a philosopher also influential in the development of such approaches to the mind.[7] Later Meerbote, Kitcher, Brook and others all developed these sorts of ideas of Kant as a progenitor to the currently popular functionalist approach to the mind. Moreover, in a way somewhat parallel to the development of this idea of Kant as a functionalist, there has also developed, but now mainly in relation to Kant's moral rather than theoretical philosophy, the idea of Kant as a "compatibilist."[8] Here Kant has been interpreted as holding views of the mind–body relation somewhat like those of Donald Davidson. Both of these views suggested a picture of Kant as a philosopher of mind, who, despite his idealism, was putting forward in a general way a view of the mind compatible with more modern "scientific" approaches. Andrew Brook, for example, who has probably made the strongest of the recent claims of the contemporaneity of Kant, sees Kant's own assumption of the unknowability of the mind "in itself" as itself just a version of the functionalist's idea that any particular functional state can be instantiated in multiple ways. In Brook's terms, Kant's skepticism here was just a recognition that one cannot infer anything substantive about the mind's material form merely from a knowledge of its function.[9]

But Kant's philosophical psychology not only theorized about the nature of cognitive activity and "content," it also, as seen in Chapter 3, theorized the necessary conditions of consciousness. Given the recent revival of interest in consciousness, it is perhaps not surprising that Kant's philosophy has been approached from this direction as well. Of the current range of analyses of the structure of consciousness, the so-called "higher-order thought" (or "HOT") conception has a distinctly Kantian ring.[10]

But beyond these factors particular to the context of philosophy of mind, there are probably other much more general ones also relevant to Kant's newfound acceptability. Among these we might count the development of forms of Kant interpretation that have freed him from the unwanted metaphysical postulation of a realm of unknowable "things in themselves," somehow hidden behind the world of empirical appearances. Kant, it is now commonly argued, did not posit the existence of two separate "worlds"; rather, he posited the difference between "two aspects" that the one world presented to the mind. Things can be considered or thought about, as they are "in themselves," or they can be known as they appear to us within the constraints imposed by the conditions of our experience and understanding.[11] Beyond this, epistemology and, in particular, philosophy of science have tended in the second half of the twentieth century to swing away from the earlier more orthodox empiricist approaches and to have become, in a more Kantian way, directed to structural conditions of knowledge. Moreover, recent historians of the growth of nineteenth-century psychology have also reasserted in importance of Kant.[12]

Anything like a survey of this complex and dynamic field is beyond the scope of this work, and in this chapter I restrict myself to the issue of the degree to which Kantian "transcendental psychology" can be seen as a forerunner of a cognitive approach to the mind, an idea forcefully argued for by Patricia Kitcher in *Kant's Transcendental Psychology*. It is against such a reading of Kant that I go on in the following chapters to situate post-Kantian idealist thought as facing the question of how such cognitive capacities might be embodied. To understand what is at stake in this reading of Kant, however, it is necessary first to contrast this somewhat radical reinterpretation against an earlier distinctly antipsychological view of Kant, a more traditional "logical" account put forward by Peter Strawson in *The Bounds of Sense*.[13]

The Unity of Apperception in Strawson's Logical Kant

In his influential interpretation of Kant, *The Bounds of Sense* (1966), Peter Strawson attempted to pan off from the gold of Kant's epistemological insights the dross of the unwanted metaphysics of "transcendental idealism." According to Strawson, Kant had made "very great and novel gains in epistemology, so great and novel that, nearly two hundred years after they were made, they have still not been fully absorbed into the philosophical consciousness" (29). What prevented these from being fully appreciated, however, was that they were presented embedded within a metaphysical framework expressed in the psychological idiom of the states and activities of "departments or faculties of the mind" (20). Within this framework Kant misconceived what was properly a conceptual investigation of the fundamental structure of concepts in terms of which we make intelligible to ourselves the idea of our experience of the world as an investigation of our "actual cognitive constitution." In short, for Strawson Kant's real insights had been hidden by his falling victim to the fallacy of psychologism.

Nowhere was the need for a depsychologizing of Kant more necessary than in the "transcendental analytic" of the first *Critique*, especially the section entitled the "Transcendental Deduction of the Categories." According to Strawson, there Kant had mixed an analytical argument "about the implications of the concept of experience in general" with a story about the production of experience (stronger in the first, A, edition of the *Critique* than in the later B edition) couched in the idiom of the psychologizing metaphysics of transcendental idealism (88). Importantly, the analytical argument concerned the concept of, not the psychological dynamics of, experience. That is, within Strawson's analytic philosophy, the object of investigation here was the *meaning* of experience, not its "mechanics."

In the transcendental deduction, Kant had famously asserted of experience that "it must be possible for the 'I think' to accompany all my representations."[14] This principle of the potentially self-conscious nature of experience is, according to Strawson, in fact part of the very concept of experience: "There is, no doubt, reason to think that there are forms of sentience which fall short of this standard [of self-consciousness]. But the fulfilment of the fundamental conditions of the possibility of self-consciousness, of self-ascription of experiences, seems to be necessary to any concept of experience which can be of interest to us, indeed to the very existence of any *concept* of experience at all."[15]

The reason that this possibility of "self-ascription" is so crucially part of the very meaning of experience is in Strawson's account centrally bound up with what he calls the "objectivity thesis." It is necessary because it is implicit in the very capacity of a subject to understand its experience as experience of an objective, external reality: we might say that for Strawson it is part of the concept of cognitive experience that it is intentional. And this in turn implies that the subject can form a separation within its experience between aspects coming from the world and aspects coming from itself: "Kant's genius nowhere shows itself more clearly than in his identification of the most fundamental of these conditions in its most general form: viz. the possibility of distinguishing between a temporal order of subjective perceptions and an order and arrangement which objects of those perceptions independently possess—a unified and enduring framework of relations between constituents of an objective world" (29).

It is essential here that the notion of experience in question be understood as that of cognitive experience: the subject of that experience must understand that in its experience it is put in touch with a world whose "order and arrangement" may be different from that found among the "parts" of that experience itself. Strawson's thought here points to the example Kant had used in the Second Analogy; while we typically experience the various parts of a spatial whole in temporal succession we still know it as spatial. We must therefore be able to distinguish between this temporal subjective order and the actual objective order of reality.

As we can see, Strawson's reading of Kant results in a conception of the nature of experience not unlike that we have seen in James's later essays on the nature of consciousness. For James we must be able to think of any mental content as entering into two different orders of relations, on the one hand, that reflecting the contingencies of the individual subject's own psychological history, on the other, that reflecting the "order and arrangement" of an independent world. And as in James, this Strawsonian Kant will be found to hold to a somewhat *direct realist* account of perceptual intentionality (27). But for Strawson, this nonsubjective ordering will be fundamentally secured at the level of the logical relations between experiential contents. This is why cognitive experience must be understood as Kant understood it, as judgmental and conceptual; and objects of experience must be understood as the "topics" of such judgments. As Strawson points out, the word "object" for Kant "carries connotations of 'objectivity.' To know something about an object, e.g. that it falls under such-and-such a general concept, is to know something that holds irrespective of the occurrence of any particular state of consciousness, irrespective of the occurrence of any particular experience of awareness of the object as falling under the general concept in question. Judgments about objects, if valid, are objectively valid, valid independently of the occurrence of the particular state of awareness, of the particular experience, which issues in the judgment" (73).

To articulate experience conceptually or judgmentally is, potentially at least, to bring it into the Sellarsian "space of reasons," into a potential web of logical relations with other confirming or falsifying judgments. Thus the unity that is at issue among our judgments is the unity that accrues to them in virtue of the fact that they are all about the one world. To express this in the Fregean idiom, that they all have as their meaning (their *Bedeutung*), "the true": As Strawson points out,

> Kant thinks of this requirement of unity and connectedness of representations as extending to the whole course of our experience. No breaks can be allowed in it. A particular unruly perception is not reckoned as a glimpse of another objective world, but is relegated to the status of subjective illusion. Similarly, if any phase of experience is to count as a phase of experience of the objective, we must be able to integrate it with other phases as part of a single unified experience *of a single objective world.* (89, emphasis added)
>
> If . . . our experience is to have for us the character of objectivity required for empirical knowledge, our "sensible representations" must contain some substitute or surrogate for awareness of the real, unknown object. This surrogate is precisely that rule-governed connectedness of our representations which is reflected in our employment of concepts of *empirical* objects conceived of as together forming *a unified natural world,* with its own order, distinct from, and controlling, the subjective order of perceptions. (91, second emphasis added)

For Strawson, what is misleading in Kant's psychologized way of expressing these points about the unity of experience, a unity that in fact is grounded in the unity of the cognized world, is that it is expressed as a unity brought about by some mental act of "synthesis." This doctrine of the "subjectivity of the source of such order in Nature as is necessary to yielding the unified objective world of our experience [is] an affirmation which parallels, on the side of the active faculty of understanding, the thesis of the subjective source of the spatial and temporal modes of relation in the passive faculty of sensibility" (96). It is this "picture of the receiving and ordering apparatus of the mind producing Nature as we know it out of the unknowable reality of things as they are in themselves" that was among the "chief obstacles to a sympathetic understanding of the *Critique*" (22). It is a picture that belongs to "the imaginary subject of transcendental psychology" and that should be abandoned in favor of a method that attempts to establish "a direct analytical connexion between the unity of consciousness and the unified objectivity of the world of our experience" (97 and 96).

Such were the main dimensions of this popular interpretation of Kantian philosophy, which seemed to successfully warn off those interested in Kant's view of the mind. Strawson's picture, however, came to be challenged, both in its interpretation of the nature of Kant's "transcendental idealism" and in its understanding of those issues dealt with in terms of a doctrine of "transcendental psychology." As mentioned, the traditional understanding of transcendental idealism as a metaphysical doctrine about a realm of unknowable "things in themselves" standing behind the realm of phenomena, the so-called "two-world" interpretation, was challenged by an interpretation, the "two-aspect" interpretation, which was, in fact, rather closer to Strawson's own views than the views he attributed to Kant. Regarding Kant's psychology, it was the development of the cognitive revolution together with a swing away from the conception of analytic philosophy as "conceptual analysis" that seemed to bring about a reinterest in Kantian "faculty psychology" and "the imaginary subject of transcendental psychology." In the next section I examine one of the most influential of these turns to the psychological Kant, that of Patricia Kitcher, and following that, criticisms of Kitcher's account by Henry Allison, who, while an adherent to the more "logical" conception of apperception doctrine, sets this within the context of the "two-aspect" interpretation of Kant.

The Unity of Apperception in Kitcher's Psychological Kant

If Strawson's reconstructed "logical Kant" can be considered as located at the "conceptual" or "logical" extreme of an axis stretching from logic to psychology, Kitcher's Kant must be located at the other pole. For Kitcher, transcendental psychology is the laudable methodological core of Kant's first *Critique,* and she affirms as central and valuable within the Kantian project all those aspects which Strawson tries to expunge from Kant. For her, the "central argumentative project of the *Critique* is the examination of cognitive faculties—sensibility, understanding, imagination, and reason—to determine which aspects of our knowledge derive from them, rather than from objects." She defends Kant's "weakly" psychologistic epistemology as dealing with the same mix of normative and empirical questions that have been productively pursued by modern cognitive scientific approaches to the mind while denying that he was guilty of the strongly psychologistic reduction of the normative to the natural.[16]

From Kitcher's point of view Kant's "transcendental psychology" is a project that can be compared to the type of "task analysis" approach of contemporary cognitive scientists such as Newell and Simon:

> Transcendental psychology investigates the faculties required for the performance of basic cognitive tasks. . . . [Kant] is totally uninterested in the actual physical or psychological embodiments of particular mental processes; the only goal is to explore the requirements of various cognitive tasks. In this respect his work is centrally in epistemology and very different from empirical psychology.
>
> Nevertheless, there is a relation between transcendental psychology and what we (but not Kant) call "empirical psychology." They are different modes of addressing a common subject matter. (25)

Transcendental psychology and empirical psychology can be considered as different approaches to "a common subject matter"—the human mind—in that transcendental or cognitive psychology approaches the mind qua *knowing* or *thinking* mind. According to Kitcher, Strawson makes the mistake of thinking that the object of transcendental psychology must be either the *empirical* mind in the sense that Kant meant this (as the, albeit impossible, scientific study of the laws of association of sensation) or the noumenal self. But there is a "third self" in Kant's account, the "I of apperception," which is a "highly abstract, functional description of a thinking self" (22). But it turns out that this third self is really only a highly abstract aspect of the phenomenal self, so Kantian transcendental psychology is in fact empirical psychology carried out at a high level of abstraction (139).

In particular, Kitcher attacks Strawson's thesis that the transcendental deduction is, or should be, about the "concept" of experience—that it is an "analytic" doctrine. First, that we can analytically unpack the components of the concept of experience in isolation of empirical study is an idea she mistrusts on the basis of Quine's classic denunciation of the analytic–synthetic distinction. Furthermore, without the strong thesis that Kant is exploring the cognitive capacities of the human mind, all that one is left with of the transcendental method is the so-called transcendental argument, a strategy rigorously criticized since the heyday of "conceptual analysis."

I will return to Kitcher's displacement of conceptual analysis by the idea of an epistemic "task analysis," as well as the issue of how this displacement actually deals with the charge of psychologistic reduction of the normative to the empirical. But first it is necessary to grasp something of the way in which Kitcher's rehabilitation of Kant as cognitive scientist is meant to work. As we will see, central to Kitcher's reconstruction is a shift in epistemic focus to perceptual knowledge, a shift that has crucial implications for the "logical" understanding of Kant's concept of apperception.

At the center of Kant's Transcendental Deduction on Kitcher's account are two task analyses: "How are we able to represent objects? How are we able to make judgments about objects?" (63). Significantly, while Strawson tends to run together the notions of representation and judgment, Kitcher separates them, making the former the more basic. Again, we will return to the question of what, for Kitcher, exactly is meant by the idea of "representation" below. For the moment it suffices to note the centrality of this notion in Kitcher's Kant, for whom knowledge is, in some way, based in the mind's ability to represent the world. As she puts it, "We can know an object only if we can represent it" (65). But Kitcher's answer to the question of the how of our representation of objects is given in terms of the idea of synthesis, the synthesis of "cognitive states" into "representations," an idea further invoked to account for the subsequent unification of representations into judgments.[17]

This account of synthesis as the means for the formation of representations is set out in opposition to various rival theories of representation formation such as the Aristotelian theory that has objects as impressing their "forms" into the mind in perceptual experience, and, especially, the modern account in terms of the notion of association of ideas. It is also what allows Kitcher to interpret Kant within the framework of cognitive science. What is synthesized in synthesis is, fundamentally, sensory "data" or "information" considered as the fundamental *content* of such cognitive states: "Synthetic connection is a relation of contentual connection" (117).

It is the focus on the priority of representation as synthesis of sensory information that in Kitcher's account centers Kantian epistemology on perceptual knowledge. That *representations* are specifically involved in perceptual knowledge can be demonstrated by the example of the Necker cube, the familiar diagram popular with Gestalt psychologists, in which a transparent cube can be read as a three-dimensional cube in two different ways. It is the fact that this diagram can be grasped in two different ways that demonstrates the mediation played in our perception by "representations," understood as definite ways of unifying information. In Kitcher's words, when we look at the Necker cube we "scan the lines and vertices, and the relations among them, and on the basis of that information, we interpret the figure" in one of two ways, each corresponding to a different internal representation (75).

It is this idea of the synthesis of sensory or informational states into unified representations that is nominated by Kitcher as the "core" of Kant's analysis, as well as its most defensible aspect:

> The core of Kant's analysis is that we cannot represent objects at all unless there is some process that can construct unified representations on the basis of the multiple contents of cognitive states occurring at different times through the mediation of different senses. Any possible explanation of object cognition must include an account of this process of connection or synthesis. Further, this process cannot be governed by the unaided law of association. Synthesis must be carried out inside the mind, so it requires some mental faculty that has the power of synthesizing. Kant labels this faculty [productive] "imagination." (81)

The paradigm of perceptual knowledge is significant in a number of ways. First, the ambiguity of the Necker cube indicates that the source of the respective representational unities cannot come from the perceived "object" itself. An object may have unity, but this unity cannot be conveyed simply via the information transmitted by the senses. Its unity cannot, as it were, be *part of* the data it gives rise to. The unity must come from elsewhere—the mind's synthetic activity. "We take in information about objects through their effects on different senses at different times. Because objects are outside the mind, we cannot explain how this information is united in a representation by appealing to objects. The only alternative is that some process within the mind—synthesis—carries out the unification" (77).

As seen, Strawson had seen the "unity" involved in the mind's activity as ultimately coming from or grounded in the world, but this postulation of the ultimately objective source of unity was linked to the idea that experience had an essentially judgmental character, the character that allowed the contents of experience to be placed within the Sellarsian space of reasons, the space, as it were, of a multiplicity of logically coherent judgments. But such a judgmental character of experience, the idea that representations must be able to be brought under concepts, is for Kitcher a further aspect of Kant's analysis that not only is independent of the core, but also is less defensible.

Having established the paradigm of synthesis in the model of perceptual representation, Kitcher can then deal with the synthesis involved in judgment in a parallel way. But Kant's treatment of judgments is in fact based on unwarranted assumptions concerning the essentially judgmental character of experience so central to Strawson's analysis. But for Kitcher, the indefensibility of these noncentral assumptions does not detract from Kant's core thesis:

> [Kant's] analysis invokes two unwarranted assumptions: Representations must be able to be brought under concepts, and there is but one rule for connecting the contents of cognitive states in a representation, that associated with the concept of an object in general. . . . These assumptions should not be granted. Even if they are rejected, however, the central portion of the analysis remains intact. We can represent objects only if the imagination has rules of synthesis for combining the multiple contents of cognitive states in a unified representation. (82)

Psychological and Logical Conceptions of the Apperceiving Subject

In Kitcher's psychological reading, the notion of apperception is now to be understood in terms of the psychological process of synthesis. For Kitcher, Kant's notion of the apperceptive unity of the mind is seen fundamentally as "an abstract account of the unity of a mind" that constitutes his response to Hume's conception of the nature of personal identity (120). The problem with Hume's characterization of the self lay not in its claim that the self is made up, as it were, of representations, but rather in the notion that such representations are merely bundled. Rather than being a bundle, the self is in Kitcher's words, a "contentually interconnected system" of such states (122). In this response, however, Kitcher's Kant implicitly accepts Hume's critique of the Cartesian idea of the mind as some type of self-perceiving mental entity. It is the conception of the self as contentually connected system of states that allows Kitcher to answer "No" to whether the self is also to be thought of as the combiner of such states. That is, for Kitcher's Kant, apperception is not an "act" of the subject but rather a "subpersonal" process that brings about a unity that *is* that subject.

On the basis of this conception of apperception Kitcher criticizes the Strawsonian account in which apperception is fundamentally a doctrine of the self's necessary reflexive "ascription" of its own mental states *to* itself. In Strawson's words, "Unity of the consciousness to which a series of experiences belong implies, then, the *possibility* of self-ascription . . . the *possibility* of consciousness, on the part of the subject, of the numerical identity of that to which those different experiences are by him ascribed."[18] Such self-ascriptive approaches confuse what Kitcher calls a "metaphysical" issue (one might say here an "ontological" one) with an epistemological one: apperception is a theory of what the subject *is*—a contentually connected system of representational states—not a theory concerning how a subject *knows* that its states are its own.[19] Kitcher dismisses the idea of self-consciousness, which she understands as a matter of the subject's consciousness of its own acts of synthesis—"synthesis watching" as she calls it (127)—as necessary for the existence of synthesis itself. Here her criticisms of "synthesis watching" have the general flavor of the criticism common to functionalists such as Daniel Dennett of the idea of a "homunculus" in the head of a subject, some subjective center of consciousness for which its mental representations represent. To avoid the spiraling proliferation of little men that this view gives rise to (spiraling because the explanatory posit of the homunculus calls forth another in the explanation of *its* operations), one must stop at the level of the connectedness of the representations themselves and avoid the idea of something *further* as responsible for the connecting.

This reading of the doctrine of apperception has drawn most criticism from other Kant interpreters, such as Robert Pippin and Henry E. Allison.[20] As we have seen for Strawson the idea of self-ascription was central to the doctrine of apperception because it was part of the very "concept" of consciousness as intentional, as being about some objective reality. To be conscious in this sense was to be able to distinguish "between a temporal order of subjective perceptions and an order and arrangement which objects of those perceptions independently possess—a unified and enduring framework of relations between constituents of an objective world."[21] Similarly for Allison, the necessary self-reflexivity of apperception is *internally* connected with the objectivity implicit in the concept of experience or knowledge:

> The thesis that we are capable of a non-empirical consciousness of our intellectual activity (a consciousness of spontaneity) is a crucial feature of Kant's doctrine of apperception. . . . It is a logical consequence of his conception of knowledge; for this conception commits him not only to the view that judgment involves a synthesizing, unifying activity, exercised upon the given by the understanding, but also that it involves a *consciousness* of this activity. In short, judgment and, therefore, experience (since it necessarily involves judgment) are inherently reflexive, self-referential activities. That

is precisely why Kant insists (B 134) that apperception involves both the synthesis of the manifold and the consciousness of this synthesis.[22]

Again like Strawson before him, Allison is concerned with the "psychologistic" implications of any interpretation of Kant that like Kitcher's interprets the unity of apperception in a fundamentally psychological way. The problem with such an account is precisely the problem that Kant had pointed to in Locke's "physiological" interpretation of mental relations, the reduction of rational to causal relations: "The central claim is that, *qua* thinkers, we must regard our reason (or understanding) as determining itself in accordance with objectively valid normative principles, that is, as autonomous, and that this is incompatible with the conception of our reasoning activity as the conditioned outcome of a causal process. Presumably, the latter is the case because, considered *qua* outcome of such a process, a belief would be something that one is caused to have by the state of one's system (plus input) and that is sufficient for denying it the status of knowledge."[23]

It is this irreducibility of the relations among mental states to causal relations that allows the thinking subject's response to the sensorily given to be a truly rational one that accords with objectively valid laws or principles. It was this that Kant had insisted on with the notion of spontaneity. This in turn means that the "apperceptive I" cannot designate the individual mind: it cannot be essentially a theory of personal (or mental) identity. Such claims "about the unity, simplicity, identity and distinctness of the 'logical subject of thought' . . . are not to be confused with synthetic propositions about a real thinking subject or substance."[24] The irreducibility renders impossible that separation of "metaphysical" from "epistemological" issues that Kitcher sees as confused within the "apperception as self-ascription" accounts:

> To begin with, since the 'I' in question is not a substance, entity or even person, but merely the subject of thought considered as such, the identity conditions of the 'I' cannot be distinguished from the conditions of its consciousness of this identity. . . . If it is to be possible for the 'I' to become conscious of its identity, the synthetic unity must not only be *in* a single consciousness, it must also be *for* that same consciousness, that is, it must be a unity for the 'I.' But since for the 'I' to take its representations as unified is precisely for it to unify them, it likewise follows that it is only insofar as the 'I' is capable of unifying its representations in a single consciousness that it is capable of becoming conscious of its own identity.[25]

As we can see in Allison's account self-consciousness is not something that can be thought of as a type of quasi-empirical introspective observation of one's own acts ("synthesis watching"). In her response to the logical reading of Allison and others, Kitcher, while acknowledging that Kant uses this terminology of the I as a "logical" or "formal" subject of thought, dismisses it as helpful on the grounds that the logical–psychological distinction, besides being foreign to Kant, is unclear: "No one believes that it is a fact of formal logic that mental states belong to a subject of thought."[26]

The dispute between Kitcher and Allison thus hangs on a claim concerning consciousness, the claim that the synthetic unification of mental representations requires that the subject be conscious of this unification—conscious of itself as responsible for the unifying. To Kitcher this smacks of the homunculus, but for Allison it seems to be a part of what is meant by the notion of the synthesis of representations. Given that the decisive issue concerns a type of consciousness, to discern what is at stake we might borrow from the conception of Kant as a theorist of consciousness.

As mentioned, Kant is now commonly taken as a precursor of a contemporary approach to consciousness known as the "higher-order thought" theory.[27] In these so-called HOT approaches, a mental state (a sensation, say) is rendered conscious by being the object of a further "higher-order" mental state. Such higher-order thought theorists (such as David Rosenthal) in contrast to "higher-order experience" theorists (such as David Armstrong) hold that the mental state rendering sensations conscious are more like thoughts than perceptions.[28] On this view, the "I" to which the experiences are attributed must be something *conceived* rather than perceived (as suggested in the "synthesis watching" thesis). That is, self-consciousness, rather than being a type of singular perception of the self, would be a conception of the self *as* a being whose judgments were logically unified. This would seem to capture Allison's claim of the "nonempirical" nature of self-consciousness. Further, this would also seem to allow for the existence of a normative dimension to such self-consciousness. One's conception of oneself as a unifier of representations might thus be a conception of a type of self to which one aspired, rather than a self that one empirically *was*. This would be a "logical" conception of the self that could perhaps be understood as *internal* to the psychological conception of mental unity that Kitcher is advocating.[29] We have seen how in his *Anthropology* it was precisely this *conceptualized* type of self-understanding that Kant stressed when he commented on the relevance of the child learning the ability to represent itself in language. But how, exactly, is this idea implicated with the nature of consciousness?

Representation and Consciousness

In Kemp-Smith's translation of Kant's *Critique of Pure Reason,* the term *Vorstellung* is rendered as "representation," *Vorstellung* being the standard philosophical German translation for "idea" or "representation" as used by English philosophers such as Locke, for whom such a notion referred to a content of consciousness. The German etymology captures this link to consciousness: the verb *vorstellen* is formed from the morphemes *stellen*, to stand, and *vor*, before, and what a *Vorstellung* was thought of as "standing before" was "the mind" or "consciousness." As such, the notion seems to come with definite "homuncular" assumptions, and ideas or representations appear to be construed as the immediate *objects* of or within consciousness, objects standing before the homuncular observer.

In contrast, the twentieth-century philosophical use of representation has tended, probably because of the predominantly logical contexts in which it has developed, to circumvent this "veil of ideas" tradition and its homuncular associations. As mentioned, Kitcher avoids the term "representation" for *Vorstellung* because she believes that for Kant not all *Vorstellungen* actually "represent." By rendering *Vorstellung* as "cognitive state," a term described as "a dummy name for a state that performs a role in cognition, but whose nature we do not understand,"[30] Kitcher seems to negate any connotations concerning any subject *before* whom such *Vorstellungen* do "stand." Furthermore, the fact that such a representation is a *state*, not a quasi-"object" of consciousness, again seems to circumvent the same homuncular idea. Qua state a representation does not need a distinct internal observer, a homunculus, for whom the representation stands as a type of internal object. A representation just *is* a state of that representing subject.

Nonetheless, Kitcher readily acknowledges that representations do represent *to* someone or something. As she notes, "In Brentano's well-known terminology, Kant is trying to understand how a mental state could be 'intentional'."[31] Congruent with this, Kitcher at times speaks of representations representing this content *to* a subject—"any cognitive state, properly so called, must either represent something to a subject or contribute to such a representation (and so participate indirectly in it)" (109)—or "for" a subject: "The question about intuitions is how they can be (or function as) representational for a subject" (113). But as she thinks of a subject as a *system* of representations, such representations would presumably be thought to exist "to" or "for" the system of which they are a part. Such a "to" or "for" does place conditions on what it is to be a representation: for example it would seem that (*pace* Tye) for Kitcher causal covariation cannot per se count as representation, as it is not per se "to" or "for" anything: "The height of the mercury in a thermometer varies regularly with the temperature, but it does not represent the temperature—at least, it does not represent the temperature *to the thermometer*" (113). We do get a sense of what this "to" or "for" means for Kitcher, however, when she qualifies this view: "If a cognitive state covaries with an external stimulus *and produces appropriate behavior,* then there is some (perhaps weak) sense in which it represents that object to its subject."[32] Would the context of function within a thermostat then allow the thermometer to "represent" temperature?[33]

As seen, functionalist accounts of the mind are often indifferent to issues of "phenomenal" consciousness. Flirting with this idea of "representation in the weak sense," which could be applied to automata, Kitcher too seems to be pursuing a notion of representation indifferent to questions of consciousness. She does raise the question of consciousness when discussing *unconscious* representations, noting Kant's rejection of Leibniz's "bloated metaphysical doctrine" that individual unconscious *"petites perceptions"* reflected the entire universe. For Kant, "unconscious, inert 'perceptions' would represent nothing to their subjects, because they would have no effects in cognitive life."[34] But here the notion of a conscious content just seems to be equivalent to that of representation per se, with its connotation of its being "contentually connected" within a system. Here "conscious" would seem to mean "access conscious" and to carry no implications of "phenomenal consciousness" at all. Thus the distinction between conscious and unconscious contents seems to collapse into that of representational and nonrepresentational *Vorstellungen.*

Such a notion of consciousness is presumably of no use for Allison's Kant in giving an account of the type of "non-empirical" consciousness implicated in *self-consciousness.* What the notion of self-consciousness would need to capture, would be something closer to *phenomenal* consciousness. Thus in Nagel's classic paper, when agonizing over the impossibility of knowing what it is like to be a bat, he seems to be alluding to the impossibility of having something like a bat's consciousness of itself.[35] But Allison's Kant would not be happy with this alone as a model for self-consciousness, as this type of self-feeling is precisely what the child, in Kant's comments in the *Anthropology,* must go beyond to become self-conscious. What self-consciousness requires is a conceptualized rather than a felt consciousness of self.

Self-Awareness and the Synthetic Capacity

For Allison's more traditional Kant to resist the functionalizing reading of Kitcher, a way is needed of spelling out the nature of the nonempirical self-consciousness implicated in the very possibility of synthesis.[36] But this problem facing Kantians, I argue in the next chapter, is not new. In fact it was the issue over which disputes broke out among followers and

critics of Kant alike in the decade after the appearance of the critical philosophy. One solution posed to the question of self-awareness was to turn,
like Fichte, to self-feeling. Before going to that period and those disputes,
however, it may be useful to see a way in which this problem is not simply
one for philosophers. It can be a problem faced by individuals in a more
direct and immediate way, individuals with particular neurological disorders, for example.

Korsakoff's psychosis, a condition commonly found as a consequence
of alcoholism and in association with a distinct pattern of neural damage
to the diencephalon or temporal lobes ("Wernicke's encephalopathy"), is
characterized by severe short-term memory loss, a tendency to "confabulation," or falsification of memory, and disorders of peripheral nerves. Israel Rosenfield, however, has contested the idea that the mental deficits of
Korsakoff patients should be thought of as a deficiency of "recent" as opposed to "long-term" memory: "The distinction between 'recent' and 'remote' memory is not about memory at all but is a distinction between different ways in which the brain structures knowledge."[37] Korsakoff
patients confabulate: asked what he did "last Saturday" a patient might
recount something that was done years ago regularly on a Saturday night.
Such confabulation is commonly interpreted as the person replacing an
impaired short-term memory (they cannot remember what occurred last
Saturday) with a distant one because their short-term is more severely impaired than their long-term or remote memory. But Rosenfield asks us to
see this in another way. Both sorts of memories can *be* memories only if
they can be related to the present. What, according to Rosenfield, is disor-
dered in Korsakoff patients is their sense of time, and importantly, their
sense of the present.[38] Without a sense of how the past relates to the present, there is no way of understanding some event presented in memory
as taking place at a particular time. "Last Saturday," for example, gains its
meaning from its relation to the present. What, it seems, underlies this dis-
ordered sense of the present, is a disorder in the mechanisms giving feedback from the body to those parts of the brain that LeDoux has termed the
"emotional brain."[39]

When brain damage destroys specific aspects of self-reference, it alters the
structure of consciousness. . . . Baud's injury [i.e., Henri Baud, the patient
described in an early classic paper on Korsakoff's psychosis] meant that he
could sustain a sense of body image for only about twenty seconds, too
brief a span for him to establish a sense of time. His self-consciousness, his
self-awareness, was thus deeply altered. His "last Saturday" was, at best,
abstract; for him, "last Saturday" floated in a timeless world not relating to
any specific day past or present.[40]

A mental representation which "floats in a timeless world" is one that cannot be synthesized with others within the Kantian "transcendental intuition" of time. But such patients not only show an inability to order their memories into some synthetic whole, a narrative, as it were, in which memories can be arranged with respect to a "now," they also show peculiarities in the subjective or phenomenal qualities of consciousness. Oliver Sacks reports asking such a patient how he felt: "I cannot say I feel ill. But I cannot say I fell well. I cannot say I feel anything at all." Did he feel *alive?* "Not really. I haven't felt alive for a very long time."[41] It is tempting to describe these reactions as those of a person for whom there is nothing in particular that it is like to be that person. A person without phenomenal consciousness.

We might, then, see Korsakoff patients as suffering from a "Kantian" type of problem: they seem incapable of one particular type of synthesis, the type required for rationally incorporating "memories" into their verbal behavior and their activities, and this problem seems to be linked to their inability to synthesize transduced feedback from their bodies into some sense of how they feel and to link such a felt representation of the self to sensory information received from the outside world.[42] They seem to exemplify the opposite condition to the prelinguistic infant. Without the use of "I," the infant, in Kant's account, has a feeling of itself, but not a concept of itself. Korsakoff patients have the use of "I" and, presumably, some concept of themselves; but this concept cannot be anchored in a *feeling* of themselves. In fact, Korsakoff patients appear to lack what Fichte, in his critical reinterpretation of Kant, had referred to as "intellectual intuition."[43]

5

The Unsayable Self-Feeling Body:
Feeling, Representation, and Reality
in Fichte's Transcendental Idealism

R ecent functionalist discussions of affect like those of Tye and Charland have considered feeling as "representational" and, specifically,
as representing states of the body. But such analyses do not seem to dis-
criminate adequately between "having" a feeling and the overtly "representational" being introspectively aware of that feeling. But, as suggested
by the phenomenon of Korsakoff's psychosis, self-feeling may be implicated in our ability to form representations in another, deeper way. Rosenfield's idea of constant somatic feedback to the limbic system is similar to
what Antonio Damasio has called "background feeling."

Background feeling is feeling we are only subtly aware of "but aware
enough to be able to report instantly on its quality," which constitutes
"our image of the body landscape when it is not shaken by emotion."[1] Indeed James alluded to what seems to be the same phenomenon. Concerning that "something with which we also have direct sensible acquaintance, and which is as fully present at any moment of consciousness in
which it *is* present, as in a whole lifetime of such moments," James, gener-
alizing from his own experience, declared that, whatever this "something" was, it was *felt* "just as the body is felt."[2] Moreover, it was even felt
in very specific regions of the body:

> In a sense, then, it may be truly said that, in one person at least, the *"Self of
> selves," when carefully examined, is found to consist mainly of the collection of
> these peculiar motions in the head or between the head and throat. . . .* But I feel

quite sure that these cephalic motions are the portions of my innermost activity of which I am *most distinctly aware*. If the dim portions which I cannot yet define should prove to be like unto these distinct portions in me, and I like other men, *it would follow that our entire feeling of spiritual activity, or what commonly passes by that name, is really a feeling of bodily activities whose exact nature is by most men overlooked.*[3]

As Anne Harrington has pointed out, John Hughlings Jackson had discussed a similar primordial and submerged sense of self in the nineteenth century with his notion of "subject consciousness":

Jackson believed that his discovery of the duality of mental operations meant that one could finally begin to make sense of a puzzling but very basic feature of human cognition: the fact that, in thinking, the individual was aware on some deep intuitive level that his thoughts belonged to him and him alone; the fact that his consciousness consisted not just of knowledge but of a knower, not just of an object but of a subject. This was simply because, Jackson argued, every thought was actually activated twice, first unconsciously and then consciously. The first, unconscious activation gave rise to that tacit awareness "deeper than knowledge" . . . that one was experiencing processes of thinking and experiencing, which collectively constituted the foundation of one's existence as a thinking self. The second, conscious activation corresponded to awareness of the content of one's thoughts and experiences in their own right. Jackson proposed to call the first, automatic half of thought *subject consciousness* (even though it was not actually conscious at all!) and the second, voluntary half of thought *object consciousness.*[4]

As Harrington points out, Jackson did not mean by "subject consciousness" anything like the introspective examination of one's mental contents:

For Jackson, *only* states of object consciousness could be contemplated or known; the knower could not make himself the object of his knowledge, since that implied the absurdity that a self could at once think and observe itself thinking. As Mercier, one of Jackson's students, would later put it, subject consciousness involved "a colouring or modification of the mental self directly, and without the intermediation of any processes of thought." . . . Individuals, in short, could become conscious of changes in their "selves" (changes in the highest levels of their nervous systems) only indirectly, only through the mediation of "symbol-images."[5]

With this, Jackson here seems to have adopted an attitude toward the nature of the mind's capacities that had been initiated by Fichte and developed by Schelling in the last years of the eighteenth century. During the first decades of the reception of Kant's critiques, issues to do with the role of self-reflexivity in relation to the mind's representational powers and the relation of both these issues to the puzzling question of consciousness had been fought out in German philosophical circles. Although Fichte is commonly thought of as having cut off Kantian idealism entirely from its residual "realist" aspects and as having relapsed in a dogmatic metaphysics of the "Absolute ego," he might indeed be more correctly understood as having rearticulated Kant's mind–world relation in interesting and fruitful ways. Asserting that the mind's representational capacities depended on a more immediate *nonrepresentational* form of self-awareness, Fichte had conceived of the latter in terms of *feeling*. And as Daniel Breazeale has argued, this move was connected to an implicit "realist" dimension to Fichte's thought.[6] But Fichte was not able to make this realist dimension explicit, and it was precisely this goal which seems to have motivated his then-follower, F. W. J. Schelling, in the years around the turn of the century.

The Development of Jena Post-Kantianism

Taking up a chair in philosophy in Jena in 1787, Karl Leonhard Reinhold was the first "Kantian" professor of philosophy and established Jena as the center of critical philosophy throughout the 1790s. While starting as an expositor of Kant, Reinhold quickly became the first of a series of internal critics of Kantianism. Believing that much of the controversy over critical philosophy had been caused by Kant's neglect of the task of clarifying his fundamental principles and concepts, Reinhold set about to find a clear and explicit starting point from which transcendental philosophy could develop. And he did this by invoking that which was a conventional starting place for early modern philosophy, the nature of consciousness.

For Reinhold what was needed was some immediately given and obvious general feature of consciousness that all conscious subjects could acknowledge, and he found this in its *representational* structure. As we have seen, the notion of *Vorstellung* was basic to Kant's philosophy, but Kant had never said explicitly what he meant by it. Reinhold's starting point was to appeal to this idea as primitive and capture it in a formula that revealed its basic structure. Critical philosophy could then proceed by a type of transcendental reflection on, and exploration of, the "internal conditions" of representation. Reinhold stated his formula for representation, his "proposition of consciousness" as: "In consciousness representation is distinguished through the subject from both object and subject and is referred to both."[7]

To be a starting point, this conception of representation had to be somehow self-evident. It is, he noted, *"immediately* drawn from consciousness; as such it is entirely *simple* and incapable of analysis . . . it admits of no explanation but is self-explanatory."[8] But this self-explanatory nature was far from obvious to his critics, of whom one of the most significant was the Humean skeptic G. E. Schultze. Schultze, in his anonymously published *Aenesidemus* (1792),[9] directed against Kantian philosophy, subjected Reinhold's "self-explanatory" concept of representation to rigorously skeptical scrutiny.[10]

It was against the background of Aenesidemus' attack on Reinhold that the next and much more significant systematization of Kant was attempted, that of Johann Gottlieb Fichte. Fichte, who succeeded Reinhold at Jena, responded to Aenesidemus' attack,[11] but he was unhappy with his predecessor's starting point of the primordiality and givenness of *representation.* For Fichte, "representation" (*Vorstellung*) presupposed a more basic structure and form of awareness, that characterizing the subject's own self-conscious relation to itself—that *self*-relation which was implicit in Reinhold's separation of the represented object *from* itself and its relating of it *to* itself. This self-relation must be a form of awareness of oneself but it cannot be a *representational knowing* of oneself. Fichte called it "intellectual intuition."[12] But intellectual intuition was also described as a form of "self-positing" in which an "Absolute I" seemed to bring itself into existence. It was with this latter idea, which seemed to combine elements of the Cartesian cogito with Kant's idea of the spontaneity of apperception, that Fichte in the first pages of his *Foundations of the Entire Science of Knowledge* of 1794, the first in his ongoing project of the *Wissenschaftslehre* ("Science of Knowledge" or "Doctrine of Science"), tried to displace Reinhold's idea of *Vorstellung* as the starting point of the Kantian system.[13]

Fichte's attempt to find a starting point in the Absolute I is complex, easily misunderstood, and all too easily parodied. The well-known passages from *Foundations* concerning the "self-positing" of this *"Absolutes Ich,"* especially when taken in isolation from his elaborations on this idea elsewhere, can give a misleading impression as to what the doctrine is fundamentally about. But taken in context Fichte's odd-sounding ideas can be seen as addressing important issues in Kantian philosophy of mind.

The Fichtean Self-Positing, Self-Intuiting I

At the start of the 1794 *Foundations of the Entire Science of Knowledge* Fichte commenced his investigation into the "primordial, absolutely unconditioned first principle of all human knowledge" by an apparent distancing of his task from the more "phenomenological" approach found in Reinhold. The first principle is "intended to express that *Act* which does not and cannot appear among the empirical states of our consciousness, but rather lies at the basis of all consciousness and alone makes it possible."[14] As he pursues his task it is clear that Fichte is attempting to reformulate Kant's transcendental unity of apperception. Moreover, like Kant, he proceeds at a level of considerable abstraction by appealing to those "laws of common logic" that must hold of the *form* of any possible objective *content* of consciousness, "laws" such as that of any thing's identity to itself ($A = A$).

In Kantian fashion such laws are held as grounded in the transcendental unity of apperception itself. This idea of a necessary connection between premiss and conclusion of a valid inference must be "*in* the self, and posited *by* the self, for it is the self which judges."[15] To the extent that an object of thought *is* the bearer of such logical form, it too is "*in* the self and posited *by* the self." So far this does not go beyond Kant, although for Kant, of course, this only applies to the form of mental contents. Experience is entirely dependent on what is passively given in experience as far as the matter or existence of what is represented is concerned. But Fichte now introduces the idea of existence in another way—that concerning the existence of the "I" itself—by appealing to the idea of Cartesian self-certainty, the certainty with which an I asserts its own existence with the proposition "I am." Thus while the proposition "I am I" may at first look like just another instance of the basic logical principle $A = A$, it is in fact quite different. With this particular content, "I," the otherwise formal truth becomes an existential or material one, since it contains within it the material, self-verifying proposition, "I am." The "I" posits itself.

The trajectory of Fichte's deduction quickly becomes apparent. From another logical proposition, that "$-A$ is not equal to A," and from the inability to derive this from $A = A$, he derives the equi-primordiality of the form of the act of "counterpositing" to that of "positing." Materially, however, the counterpositing of "$-A$" depends on the positing of A. But we have seen from the discussion of the "I am" that the only thing posited "absolutely" is the I itself; that is, while any other "A" needs to be posited, by the I, the I posits itself. But this means that the positing of the self's opposite, the not-self ("$-I$"), will be *materially* dependent on the self-positing of the Absolute I itself.

With this Fichte indeed seems to have realized Strawson's worst nightmares about the consequences implicit in Kant's transcendental psychology with its confusion of questions of the *formal* conditions of objective experience with a story about the actual mind's mental *acts*. And this general assessment of Fichte seems confirmed when one considers the other well-known feature of German idealism that he introduced: the elimination of the Kantian notion of the "thing in itself." This was precisely the notion with which Kant, albeit inconsistently, had tried to hold onto the idea of the limits of the I's spontaneous acts of positing, limits imposed by the world itself. Without these limits we seem to risk what John McDowell has described as a conception of thought as a "frictionless spinning in a void."[16] Fichte seems to live up to all the dire consequences that critics have seen as implicit in Kantianism.

A more sympathetic reading of Fichte, however, is available when one puts these well-known passages in the context of some less well-known others that reveal Fichte as working with a much more experience-based approach, somewhat akin to Reinhold. Seen in this light we might see Fichte as a serious developer of the Kantian theory of mind. These passages include the methodological comments from the "Second Introduction to the Science of Knowledge" of 1787 in which he uses the terminology of "intellectual intuition" (a notion not used in the 1794 *Foundations* but used in his *Aenesidemus* review) as well as his discussion of feeling as the phenomenological analogy of the "check" to the self's activities in "Part III" of the *Foundations*, which examines issues of practical rather than theoretical reason.

In the "Second Introduction" Fichte approaches the question of the self-positing I in terms of the concept of "intellectual intuition," justifying his method in terms of a move typical of early modern philosophy, the recourse to an investigation of the nature of consciousness: "[Intellectual intuition] is the immediate consciousness that I act, and what I enact: it is that whereby I know something because I do it. We cannot prove from concepts that this power of intellectual intuition exists, nor evolve from them what it may be. Everyone must discover it immediately in himself, or he will never make its acquaintance."[17]

Fichte takes the phrase "intellectual intuition" from Kant—somewhat provocatively because Kant in the *Critique of Pure Reason* had used it to characterize the very antithesis of the discursive intelligence of finite human beings. For Kant, intellectual intuition would be what characterized God's thought—a type of thinking that not only would be capable of a direct "intuition" of things in themselves, unmediated by the necessity of empirical experience, but also would be *creative*, giving rise to the very existence of those things thought.[18] Fichte goes about defending his use of the term by rejecting one part of the Kantian scaffolding, a part that had been criticized as contradictory: the idea of the supersensuous "thing in itself." Without the thing in itself, Kant's intellectual intuition becomes a "wraith which fades in our grasp when we try to think it."[19] There are no "things in themselves" which *could* be known or created by merely thinking about them. Moreover, intellectual intuition is not the intuition of a "thing" or object at all but an action. In intellectual intuition, the subject is aware of itself as subject, and as a subject that is essentially free activity (the "spontaneity" of Kantian apperception) rather than a thing, a *Tathandlung* (Fichte's neologism—literally, "fact-act") rather than a *Tatsache* (thing or fact). We can now see how intellectual intuition is a type of self-*positing:* without the awareness there is no self or subject because this subject simply *is* this necessarily self-aware activity.

With such a critique of the "thing in itself" it might be thought that Fichte has simply dug a deeper hole for himself. By discarding the thing in itself he seems to have abandoned that residual "realist" element that, as ill-fitting in the system as it may have been, lent the Kantian system intelligibility as that which limited the I's spontaneity. Moreover, the subjective analog of the thing in itself for Kant was the passively received sensory content of representations, the only indicator for a discursive intelligence of the actual existence of the objects it conceptualized. This all seems to give Kant's view a realistic dimension. As we have seen, Kant *does seem* to ground sensation in the alterable state of the physical organism that is open to the input of the world, a world of (albeit unknowable) "things in themselves." In this his view looks something like that of Tye, who wants to make sensation a type of natural sign.[20] Thus the modern view of a "physiological" basis to mental functioning is not absent from Kant's view, it just does not play the same epistemological role that it had played in Locke—the role of guaranteeing the objectivity of ideas. And, so the skeptics might say, with Fichte's relinquishment of the thing in itself, all reality and constraint *is* surely swept aside, to be replaced by the magical, mad fantasy of an all powerful, all creative *Ich.*

This familiar story ignores, however, one of Fichte's central ideas, which is closely tied to that of intellectual intuition. It ignores the fact that Fichte had not abandoned the idea of a "limit" to the I's self-positing, but had rather simply relocated it away from the, to him, indefensible idea of the thing in itself. It is thus that in Part 2 of the *Foundations of the Entire Science of Knowledge,* "The Foundation of Theoretical Knowledge," he talks of the "check" (*Anstoss*) that the activity of self-positing must meet, a check with which he means to reintroduce a "realist" element to the transcendental picture.

> It will at once be apparent that this mode of explanation is a realistic one; only it rests upon a realism far more abstract than any put forward earlier; for it presupposes neither a not-self present apart from the self, nor even a determination present within the self, but merely the requirement for a determination to be undertaken within it by the self as such, or the *mere determinability* of the self. (189–90)

The Doctrine of Science is therefore *realistic*. It shows that the conscious-ness of finite creatures is utterly inexplicable, save on the presumption of a force existing independently of them, and wholly opposed to them, on which they are dependent in respect of their empirical existence. . . . Notwithstanding its realism, however, this science is not transcendent, but remains in its innermost depths *transcendental*. It accounts for all conscious-ness, indeed, by reference to a thing that is present independently of any consciousness, but it does not forget that, even in the course of this expla-nation, it governs itself by its own laws, and that, in course of reflecting on this, the independent factor again becomes a product of its own power of thought, and thus something dependent on the self, insofar as it is to exist for the self (in the concept thereof). (246–47)

This realistic dimension of the *Wissenschaftslehre* of the 1790s, character-ized by Fichte's description of it as a "critical idealism, which might also be described as a real-idealism or an ideal-realism" (247), is related to the idea that the "Absolute I" is only one aspect of the subject, an "infinite" aspect that must nevertheless be accommodated within the "finite" self. Well into part III of the *Foundations* Fichte notes that "here the meaning of the principle, *the self posits itself absolutely*, first becomes wholly clear" (244). He then goes on to invoke the relation of the Absolute I to the actual, finite conscious subject:

There is no reference at all [in the absolutely self-positing subject] to the self given in actual consciousness; for the latter is never absolute, its state being invariably based, either mediately or immediately, upon something outside the self. We are speaking, rather, of an idea of the self which must neces-sarily underlie its infinite practical demand [*Forderung*], though it is inac-cessible to our consciousness, and so can never appear immediately therein (though it may, of course, mediately, in philosophical reflection). (244)

From this it is not so clear that Fichte is caught up in the type of confusion of which Strawson had accused Kant, as had first seemed the case. With his idea of the absolute I as "an ideal of the self which must necessarily underlie its infinite practical demand" we might see Fichte as alluding to a notion that I broached in Chapter 4 as a "psychologically real ideal." Fichte had started his system with what looked like a purely *formal* ap-proach to the contents of consciousness, an approach like Strawson's anti-psychologistic transcendental exploration of the *concept* of conscious ex-perience. But this is not how Fichte had, apparently, intended his opening analysis to be read, and the "Second Introduction," with its starting point in "intellectual intuition" and its relation to consciousness might be read as an attempt to clarify this. There Fichte notes that intellectual intuition never occurs in isolation, as a complete act of consciousness; any more than sensory intuition occurs singly or renders consciousness complete; for both must be *brought under concepts*. Nor is this all, indeed, for intellectual intuition is also constantly conjoined with an intuition of *sense*. I cannot find myself in action without discovering an object on which I act, in the form of a conceptualized sensory intuition; without projecting a picture, no less conceptual, of what I wish to bring about. For how do I know what I seek to accomplish, and how could I know this, without having an immediate regard to myself, in projecting the target-concept as an act?—Only this whole set of circumstances, in uniting the given manifold, completes the sphere of consciousness. It is only the concepts of object and goal that I come to be aware of, however, not the two intuitions that underlie them. (38–39)

Like Husserl a century later, Fichte seems to be saying that to be conscious is to be conscious of *something*. Moreover, such a something must be determinate and hence conceptualized. This idea is of course just another way of expressing the idea that the I's self-positing or self-consciousness is always "mediated" by some object of consciousness, a counterposited "not-self." But the self's positing of its object, the non-self, is at the same time a positing of itself as finite, and limited by this non-self. That is, qua consciousness of an object, I am "determined" by that object, despite the fact that the object as posited is a product of my activity.

It is this idea of mediation that Fichte retains of Reinhold's representationalism. The contents of consciousness are always conceptualized or "represented" objects—both as the known objects of theoretical reason and those "objects" (objectives or intended outcomes) at which our voluntary actions aim in practical reason. And as my self-positing or intellectual intuition is always through these counterposits, it is therefore never immediate, nor absolute, as self-identity logically must be. This mediating limitation is thus an "alien element" that "stands in conflict with the self's endeavor to be absolutely identical," that is, to be a coherent object of thought (233–34). In this sense then, self-positing is not a static characteristic, or actual achievement of the self, but rather a tendency, a striving "to maintain itself in this condition," the condition of self-identity (233).

In the context of his dismissal of the Kantian thing in itself in the "Second Introduction," Fichte makes what at first seems to be a strangely empiricist claim concerning existence: "we recognize [the thing in itself] to be the uttermost perversion of reason, and a concept perfectly absurd; all existence, for us, is necessarily *sensory* in character, for we first derive the entire concept of existence from the form of sensibility; and are thus completely protected against the claim to any connection with the thing-in-itself" (45–46). If this is the case, intellectual intuition, this type of never-achievable striving for self-identity in the face of a constantly operating "check" on this striving, must have some experiential and sensory, but nonrepresentational, form of presentation within the self. And so it does. Fichte here invokes an analogy with the participation of sensation within intuition: anticipating James's critique, Fichte declares that sensation never appears within consciousness as such, as an "isolated presentation," but always in some conceptualized form, as posited "intuition," or, we would say, perception. Nevertheless the philosopher can arrive at the "knowledge and isolated presentation" of it by "inference from the obvious facts of consciousness" and can arrive at a similar presentation of intellectual intuition by the same means (39). This perspective seems distant from the opening sections of the *Foundations of the Entire Science of Knowledge*, but in Part III of the work Fichte explicitly discusses the nonrepresentational experiential form that "checked striving" takes: it is *feeling (Gefühl)*.

Body and Mind in Kant and Fichte

Let us pause here to reassess what is at stake in Fichte's reinterpretation of Kantian idealism. As seen, Kant seemed to regard his idealist account of the mind, centering on the "spontaneity" of apperception (and the analogous "autonomy" of practical reason), as perfectly compatible with the recognition that deterministic "physiological" processes govern the operations of the mind's worldly substrate, the brain. Sensation is sometimes dealt with, I have suggested, in a psychophysiological, almost Fechnerian, way, in which the intensive, qualitative characteristics of raw, unschematized sensation are regarded as the subjective side of neurophysiological processes, and so subject to deterministic laws such as those of association. The independence of the properly *cognitive* functioning of the mental from processes of this sort by virtue of the operation of "abstraction" seems to give a picture analogous to the modern "functionalist" separation of cognitive processes from those physiological processes in which they are implemented. Thus while the mind's processes are embodied in the physical processes of the brain, they can be still regarded as having a certain organizational independence from the physical characteristics of the medium.

At another level, however, Kant also seems to have thought that the link to the world via the mechanisms of sensation, providing the "content" of those intuitions synthesized in judgments, was what ensured that the thing known about contributed to the processes in which it came to be known. Without this the mind would be simply "spontaneous," there would be nothing to stop thoughts from "spinning freely in the void." But while it is clear that some account of the mind's constraint by the world needs to be given, the particular solution that Kant offered here has always had its critics. Among the more recent has been Wilfrid Sellars with his criticism of the "myth of the given";[21] among the earliest was Fichte, with his joint rejection of "intuition" as a given content, isolable and independent from the mind's active "positing," and the concomitant idea of a supersensible "thing in itself," unknowable but causally responsible for such passively received intuition.

What allowed Fichte, in spite of his criticisms of Kant, to think of himself as a *follower* of Kant was the fact that the origins of this criticism existed in Kant's philosophy itself. In short, Kant was inconsistent. Kant wanted the givens of sensation to constrain judgment. But Kant himself had criticized Locke's "physiological" substitution of causal relations for justificatory ones—the central point being nicely phrased in McDowell's quip that the best *causation* can contribute here is not justification so much as "exculpation."[22] For Kant, the *most* the causal impact on the world could deliver was "sensation." But considered as isolated from the mind's activities, sensation was unconscious: without "extensive" relations conditional upon schematization and conceptualization, it could not play a role in justification, it could never be placed within the "space of reasons." It was, therefore, really no more than feeling.

But in Fichte's new scenario, making explicit what had only been implicit in Kant, there is no reason to think that experiencing, judging, thinking, and so on do not still *depend* on the processes of the body—the excitations of the sensory surfaces, the transmission of impulses in nerves, and so on. In this scenario, "intuitions" may still *need* sensations for their content, although there is no longer any "given" point-to-point dependencies between the *cognitive* contents of intuitions and the underlying physical processes in which they are embodied. (In the modern jargon, there may be "token" identities between cognitive experience and its physiological substratum but no "type" identities.) Something like this seems to be contained in Fichte's assertion that "in respect of its existence the self is dependent; but in the determinations of this its existence it is absolutely independent."[23] This disruption of this causal link between mind and world centering on sensation has often met with the accusation of a view of the mind disastrously cut off from the determinations of the world. Such a view, however, may be overhasty.

With Fichte's more "realist" notion of that "check" that the mind's outward striving necessarily meets we can see the mind's necessary limitation as simply having been *relocated*. The positing is not checked by any "thing in itself," an unknowable analog of the posited appearance standing somehow behind that appearance. But neither is it "checked" by the posited not-self, the known object. We can never say what the I's striving is checked by. Striving and check are always conjoined and are the preconditions of all positing and so cannot be explained by what is posited. Fichte's objection to this latter, somewhat commonsense view, is nicely summed up when he notes that "if we fancy some intelligent being outside the self, observing the latter in these two different situations, then *for such a being,* the self will appear restricted, its forces rebuffed, as we take to be the case, for example, in the physical world" (234). That is, that standard realist viewpoint (what McDowell calls the "sideways-on view") presupposes a possible "intelligent being" for whom that could be a cognitively relevant "view," and so surreptitiously presupposes the positing performed by that imagined being. "But the intelligence positing this restriction is not to be some being outside the self, but the latter itself" (234).

As Daniel Breazeale has pointed out, for Fichte, the subjective presentation of such a checked striving is dealt with in Section 3 of the *Foundations of the Entire Science of Knowledge* in terms of *feelings.* Feelings are *"what* we actually experience in the *Anstoss."*[24] They are the index of a constraint or check to the spontaneous activity of the I, its "checked striving," but they are not representations of the I's posited "objects." It is this alternative "realism" that gives Fichte a different way to think about the mind–world relation.

In the traditional picture found in Locke, a sensation qua "simple idea" is treated representationally and as a component or element of a more complex or compound representation (a complex idea) that acts as a type of epistemological interface between the mind and world. (In contemporary accounts, the "representational" nature of such a simple idea would be reduced to "causal covariance.") As in Reinhold's view, representations are seen as facing both ways: they are representations *of* the object and they are *for* the subject because they are somehow *in* or *of* the mind of that subject. The mind thus meets reality in them via their capacity to represent. Kant's view is not consistently "representational" in this sense, but the traditional view seems implicit in his well-known formula: "Thoughts without content are empty, intuitions without concepts are blind."[25] In this regard, Fichte's reworking of the phrase is revealing: "Intuition *sees,* but is *empty;* feeling *relates to reality,* but is *blind."*[26]

It may seem strange to say that Fichte has shifted the relevant nexus between mind and world to a more fundamental level, the level of nonrepresentational feelings rather than representations because feelings are taken by him as the means by which the mind is nonrepresentationally aware of just itself. But it would seem to be precisely in relation to its worldly, "real" aspect, rather than its cognitive "ideal" or "representational" aspect, that the mind feels. These feelings are an index of an embodied active self's checked strivings. Without this practical movement into the world, the world would not, as it were, strike back. Furthermore, for Fichte, feeling provokes reflection by the self, which results in the self positing an object (a non-self) onto which the feeling is projected to give it determination.

The distinctive phenomenological qualities of secondary properties provide an example of this. Along with Kant (and most other moderns), Fichte thinks of color or taste, for example, as "manifestly something purely subjective." "Anything sweet or sour, or red or yellow, is absolutely incapable of being described, and can only be felt, nor can it be communicated by any description to someone else. . . . All that can be said is that *the sensation of bitter, sweet, etc., is in me,* and nothing more. . . . Such purely subjective relationships to feeling are the source of all our knowledge; without feeling, there can be no representation at all of an external thing" (274–75). But although the source of knowledge, such feelings are themselves not a form of knowledge—they are not representations. Rather, they are the occasions for reflective positing, the self's formation of representations. "This determination *of yourself* you now carry over at once to something *outside you;* what is actually an accident of your self, you transform into the accident of a thing required to be external to you . . . *a matter that must be extended in space, and occupy the latter*" (275). This matter is, of course, a posit "framed or thought through productive imagination," but it is posited *as* objective. But "without a subjective property to be transferred to it, such a matter simply does not exist for you, and is thus nothing more for you than the bearer you need for the subjective property that is to be carried over from yourself" (275).

One wants to say here something like "for Fichte, in feeling, the mind is aware of itself *as embodied* and as interacting with a real world of objects," but to say this is again to "fancy some intelligent being outside the self" who is able to observe, as it were, the two sides of the relata: the feeling itself and its concomitant brain process. But such an intelligence would "posit" that which is felt in feeling on the one hand and the complicated material processes on the other, and this is precisely the "transcendent realism" to which Fichte, in true Kantian style, is opposed. Moreover, we might see his opposition here as an opposition not to the idea of an embodied mind, but rather to the very unembodied idea of the mind presupposed as the mind of that surreptitious "intelligent being outside the self" on which the transcendent realist's position depends.

On such a reading Fichte has a deeply "realistic" view of the mind, but it is one that has to remain unspoken and unspeakable. On the analogy with Wittgenstein, it might be said that for Fichte realism can be "shown" but not "said." Thus Fichte refers to an "original interaction between the self and some other thing outside it, *of which nothing more can be said*, save that it must be utterly opposed to the self" (246, emphasis added). Later Schelling, reflecting on the limits of Fichte's approach, describes self-consciousness as "the lamp of the whole system of knowledge," which, however, "casts its light ahead only, not behind"[27] It is this potential implicit in Fichte's view that Schelling sets out to release in the next significant phase of Jena post-Kantian idealism. Before turning to this, however, I wish to explore further some of the consequences of this shift to the primacy of feeling in Fichte.

Feeling and Sensation in Kant and Fichte

It will perhaps help in assessing the relevance of feeling in Fichte to contrast his views with those of Kant. Kant, as is well known, was no particular friend of feeling; when it appears in his philosophy it almost invariably plays a negative role.[28] Take, for example, his treatment of feeling in one realm within which there was a tendency to attribute to it a positive value: aesthetics.

The whole edifice of aesthetic judgment as developed in the first part of Kant's *Critique of Judgment* is built on the contrast between that which is beautiful and that which is merely "agreeable" or "charming." While the value of the agreeable or charming is entirely "subjective," in the sense of relative to the viewer, that of the beautiful is "objective" and normative for *all* viewers. Although all subjects may not judge the same things beautiful, they *should* do so: if a person "proclaims something to be beautiful, then he requires the same liking from others; he then judges not just for himself but for everyone, and speaks of beauty as if it were a property of things."[29]

Such universality can only come from the *form* of the representations involved. Pleasures issuing from the sensory appreciation of objects via the *contents* of their representations cannot be regarded as the bases of aesthetic judgment. If a person says "that canary wine is agreeable he is quite content if someone else corrects his terms and reminds him to say instead: It is agreeable to *me*. This holds moreover not only for taste of the tongue, palate, and throat, but also for what may be agreeable to any one's eyes and ears."[30]

Kant's position here is complex, for the "feelings" of pleasure taken in qualities such as taste or color are doubly subjective. First of all, using his own version of the primary/secondary quality distinction, Kant considers taste and color as *themselves* "subjective" qualities. Consider, first, the following passage from the *Critique of Pure Reason*, where the traditional idea of a "secondary quality" is translated into the framework of transcendental idealism:

> The taste of a wine does not belong to the objective determinations of the wine, not even if by the wine as an object we mean the wine as appearance, but to the special constitution of sense in the subject that tastes it. Colours are not properties of the bodies to the intuition of which they are attached, but only modifications of the sense of sight, which is affected in a certain manner by light. . . . Taste and colours are not necessary conditions under which alone objects can be for us objects of the senses. They are connected with the appearances only as effects accidentally added by the particular constitution of the sense organs.[31]

Reverting to the traditional terminology we here get the view that taste and color are not primary qualities ("objective determinations"), but rather secondary ones—simply the "effects" of the perceived objects in the subject that are "added" to or projected upon the object (the appearance) constituted by those primary qualities (objective determinations). But, invoking a distinction between "objective" and "subjective" sensations, Kant further differentiates between color and taste themselves and "appreciative" qualities "agreeableness" and "pleasantness," which we commonly attribute *to* those former qualities (that is, between secondary and what are sometimes called "tertiary" properties). Thus, in his discussion of the subjectivity of agreeableness in *Critique of Judgment* he notes that: "The green color of meadows belongs to *objective* sensation, i.e., to the perception of an object of sense; but the color's agreeableness belongs to *subjective* sensation, to feeling, *through which no object is represented*, but through which the object is regarded as an object of our liking (which is not a cognition [*Erkenntnis*] of it)."[32]

Clearly, given his views on the "secondary qualities" of color and taste from the first *Critique*, the "objectivity" of the color appealed to here is in some sense illusory: the color is only a modification of the "sense of sight, which is affected in a certain manner by light," not the objective appearance of the meadow itself. How then does color differ from "agreeableness" as both seem to be somehow projected onto or added to an objective appearance? The crucial thing seems to be the issue of interestedness and whether or not the sensation satisfies some felt lack: "Both the agreeable and the good refer to our power of desire and hence carry a liking with them."[33] So when I feel the taste of a wine to be "agreeable," it is not the wine (that is, its "objective" appearance) that is presented through the feeling, but rather my body itself, although this presentation is noncognitive. The wine, we might say, only features as an index of what my body lacks or wants.

Why does interestedness play this dichotomizing role? There are, after all, alternative ways of capturing the distinction between secondary and appreciative or "tertiary" qualities—for example, making the difference one of degree. Given what we share with most other humans in terms of the physiology of our visual systems, we are probably more likely to agree about colors than many appreciative qualities, but it is easy to overlook the degree of commonality found in these latter ones: it is probably uncommon to find humans for whom the smell of fresh feces is "agreeable," and there is most likely an explanation for this in terms of the evolutionary history of our perceptual senses.

What seems to be behind Kant's thought here of the "subjective sensations" or feelings as being somehow primarily (but noncognitively) "about" one's body is the idea that their significance lies in the role they play in action and practical reason, their role in the identification and satisfaction of bodily needs.[34] But given the effort within Jena Kantianism to overcome the dichotomization of theoretical and practical reason, it would have been unlikely for any such hard-and-fast distinction between "secondary" and "tertiary" qualities to survive. That is, within such an attempt to systematize Kant, one might expect to see the reincorporation of "feelings" into the operations of reason. And so one does, in Fichte's notion of intellectual intuition.

Fichte gives a clear sense of the connection of intellectual intuition with the feeling of *agency* in the "Second Introduction" of the *Wissenschaftslehre.* "I cannot take a step, move hand or foot, without an intellectual intuition of my self-consciousness in these acts; only so do I know that *I* do it, only so do I distinguish my action, and myself therein, from the object of action before me. Whosoever ascribes an activity to himself, appeals to this intuition. The source of life is contained therein, and without it there is death." This is not to say that the sense of agency he appeals to is exclusively that of bodily actions. It also includes the sense of our agency we have in voluntarily directing our own thinking: "I propose to think of some determinate thing or other, and the required thought ensues; I propose to do some determinate thing or other, and the representation of its occurrence ensues."[35] It is sequences of mental content such as these that present the philosopher with the "facts of consciousness" from which the existence of intellectual intuition can be inferred, however, the facts of consciousness to be perceived rely on the appeal to a practical attitude with respect to *both* thought and outer action:

Were I to view it by the laws of merely sensory consciousness, it would contain no more than has just been given, a sequence of particular representations; this sequence in the time-series is all I would be conscious of, and all

I could assert. I could say only: I know that the representation of that particular thought, with the indication that it was to occur, was immediately
followed in time by a representation of the same thought, with the indication that it actually was occurring; that representation of that particular
phenomenon, as one that was to happen, was immediately followed by its
representation as really taking place; I could not, however, enunciate the utterly different proposition, that the first representation contains the *real
ground* of the second; that through thinking the first, the second *came about*
for me. I remain merely passive, the inert stage on which representations
succeed one another, not the active principle which might bring them
forth.[36]

For Fichte the basis of this second proposition does not lie in the mere sensory ingredients as they occur in time; rather, it lies in "an intuition of
sheer activity, not static but dynamic; not a matter of existence, but of
life"(39–40).

In Part III of the 1794 *Foundations of the Entire Science of Knowledge* it becomes clear why intellectual intuition is so intimately connected with feelings of bodily agency. As we have seen, intellectual intuition is always a
striving, a striving to overcome some particular encountered "check," and
in consequence to this, some "posited" non-self opposed to the self. This
striving is always manifested in *feeling,* the conscious concomitant of the
original limitation of the check, and in turn it provokes the reflective
positing of some object as the origin of the limitation and feeling. In this
way striving and reflection are reciprocally related: "All reflection is based
on the striving, and in the absence of striving there can be no reflection.—
Conversely, in the absence of reflection, there is no striving *for the self,* and
so no striving *of the self,* and no self whatever"(258). The striving itself can
be reflected on and so posited as an object of consciousness, and as such it
appears as *drive (Trieb).* Fichte thus describes the "circuit of the self's functions": "No restriction, no drive . . . no drive, no reflection . . . no reflection, no drive, and no limitation and nothing that it limits"(258).

This determination of the self as "driven by an impulse lying within itself" will in the first instance have no object, but will merely be *felt* as a
feeling of being driven out "towards something totally unknown." This is
the feeling of "longing" through which a drive is revealed as "a *need,* a *dis-
comfort,* a *void,* which seeks satisfaction, but does not say from whence," a
longing that is "the *original, wholly independent manifestation* of the striving
that lies in the self" (265, 267).[37]

We might note here some of the implications of the Fichtean position
versus Kant's views on the role played by "interestedness" to the mind's
representational powers. For Fichte the primordiality of the finite self's

striving becomes translated in a pragmatist-like way into the idea that *all* "positing" and representing of an external realm presupposes the self's practical, outwardly "driven" relation to the world, the drive that is manifested in the feeling of longing. It is only against the background of the outwardly driven activity manifested internally as "longing" that the self *can* feel itself to be "restricted" and hence feel the need to posit an object outside of itself as responsible for the restricting. Thus Fichte insists: "This longing is of importance, not only for the practical, but for the entire Science of Knowledge. Only thereby is the self *in itself*—driven *out of itself*: only thereby is an *external world* revealed *within it*."[38]

Fichte's *Wissenschaftslehre* is redolent with intimations of the body and its subjective states, which stand in some reciprocal determination with the world in which it is active, states somehow lying immediately "behind" the mind and its self-consciousness, states that are immediately and nonrepresentationally manifest in feeling. But at the same this cannot be *said*. To refer them to the body is, by necessity, to *posit* the body as the "non-self," which is the *cause* of these feelings, to posit "drives" as that which are behind its striving. But this position cannot capture the relation involved. The difficulty here is for the subject to break out of the dichotomous opposition between the immediate and speechless *felt relation* to one's embodied self and the "third-person" representational relation to one's objectified body.

This muted return to the body and its states found in the Jena writings of Fichte in the last decade of the eighteenth century was taking place in an atmosphere filled with an intense interest in the nature of feeling characteristic of the early romantic movement. This period was also marked by the development of the biological and medical sciences, and, importantly, the "rediscovery" of the philosophy of Spinoza. In the wake of Fichte's reinterpretation of Kant, a much more explicitly bodily based conception of the self was developed by one of Fichte's enthusiastic followers, Friedrich Wilhelm Joseph Schelling.

6

The Feeling and Representing Organism: Schelling, Transcendental Idealism, and *Naturphilosophie*

In the "First Introduction to the *Wissenschaftslehre*," Fichte had claimed that there are only two possible philosophical systems, idealism and dogmatism, each resulting when different elements or aspects of experience are taken as the *grounds* of that experience. These elements, always conjoined in experience but separable in thought through abstraction, are "*the thing*, which must be determined independently of our freedom and to which our knowledge must conform, and *the intelligence*, which must know."[1] The philosopher who abstracts from the former and retains intelligence is the idealist, the one who leaves out the latter and retains "a thing-in-itself . . . abstracted from the fact that it occurs in experience," is the dogmatist. Thus each will have a different account of the production of the objective representations of experience: In Fichte's words, "According to the former system, the representations accompanied by the feeling of necessity are products of the intelligence which must be presupposed in their explanation; according to the latter, they are products of a thing-in-itself which must be assumed to precede them" (9).

That phenomenological dimension to Fichte's thought, which means that the "finite rational being has nothing beyond experience" (8), implies that neither dogmatist nor idealist can refute the other. Justification for either view cannot be found in experience "for their quarrel is about the first principle, which admits of no derivation from anything beyond it." All either can do in an effort to refute the opponent is to deny what the other finds essential to experience, "and they have no point at all in common from which they could arrive at mutual understanding and unity. Even if they appear to agree about the words in a sentence, each takes them in a different sense" (12).

Thus from a "speculative point of view" idealism and realism seem of equal value, and so preference for one over the other must be grounded in something else. This, asserted Fichte, is in fact, simply "*inclination* and *interest:*" "What sort of philosophy one chooses depends, therefore, on what sort of man one is" (15, 16). Such a quasi-existential grounding of philosophical preference is tied to the fact that the different answers given to the question of the grounding of experience have a crucial implication for how the philosopher represents him- or herself. The idealist insists on freedom, but everything the idealist explains by the free activity of the intelligence, the dogmatist explains by the brute action of the thing in itself. The idealist "must represent himself as free," but "every consistent dogmatist is necessarily a fatalist" (14, 13).

Schelling's Integration of Idealism and Realism in the *System of Transcendental Idealism*

Fichte's antithetical philosophical systems reappear in a transformed manner in Schelling's *System of Transcendental Idealism* of 1800. In the Foreword, Schelling describes his purpose as to "enlarge" Fichte's transcendental idealism "into what it really should be, namely a system of all knowledge."[2] Like Fichte, Schelling makes much of the idea that a philosophy should be presented in a coherent, self-contained *system*. The systematic approach will involve "presenting every part of philosophy in a single continuum" with "no necessary intervening step . . . omitted." But in an apparent allusion to Spinoza, for Fichte the most consistent dogmatist, Schelling then states his actual motive for "devoting particular care to the depiction of this coherence" as his long-held belief in the "parallelism" (*Parallelismus*) of nature with the intellect.[3]

In the years preceding, Schelling, while working from the basis of Fichtean idealism, had nevertheless attempted to somehow link it to a "philosophy of nature," a philosophy that from Fichte's point of view exemplified the dogmatic inverse of transcendental philosophy. Immediately after the *System* of 1800, starting with the work *Exposition of My System of Philosophy* of 1801 and lasting until 1804, Schelling was to embark on a Spinoza-style attempt to link intelligence and extended matter within his "identity philosophy" and to become more explicitly critical of Fichte.[4] In the somewhat transitional work of 1800, however, Schelling seems to want to cash out those implicitly realist dimensions of the *Wis-*

senschaftslehre, and he attempts to do this by somehow incorporating Fichte's opposition of idealism and dogmatism back within the structure of transcendental idealism itself.

The complete demonstration of the parallelism of nature and of the intellect, he contends, is not possible from within either transcendental philosophy or the philosophy of nature considered independently: "*Both sciences together are alone able to do it, though on that very account the two must forever be opposed to one another, and can never merge into one.*"[5] Although Schelling is neither clear nor consistent as to precisely how existing Fichtean transcendental philosophy and philosophy of nature can be reconciled into a system, that of his own "enlarged" system of transcendental idealism, one basic idea seems clear enough: both Fichtean transcendental idealism and its complimentary nature philosophy (or "realism," "materialism," or "dogmatism," as he variously calls it) present philosophical perspectives onto the one absolute reality and, while opposed, are nevertheless *mutually presupposing*. The expanded system of transcendental idealism will somehow show how it is that these presuppose each other, but will do it without reducing the perspective of either to that of the other or the perspectives of both to that of some third. This mutually presupposing character of each of the two sciences will give the system itself a circular structure: regardless of whether one starts with the approach of Fichtean idealism or nature philosophy, one will at some stage be carried over into the perspective of the other, and, from there, eventually be in turn carried back to one's original perspective. (The parallel lines, it would seem, mysteriously join up!) A starting point must be chosen, however, and Schelling chooses that of Fichte's Kantian science, the *Wissenschaftslehre*.

Commenting on this Fichtean project, Schelling describes it in fundamentally *epistemological* terms. Here the *reality* (objectivity) as well as *systematicity* in our knowledge is assumed.[6] But such a system of knowledge must contain the "ground of its subsistence" *within* itself, that is, the "principle" of such a system must somehow lie *within* knowledge itself. What this principle is, Schelling answers in a Cartesian way: it is the principle of self-consciousness, the self's immediate awareness of itself. However, he quickly goes on to separate his own (and Fichte's) *epistemological* way of understanding this principle from an *ontological* way that was effectively that of Descartes. While Descartes had argued to an ontological conclusion, certain knowledge of the mind's existence, Schelling points out that transcendental idealism "has *knowledge*, not *being* as its object; and that its principle, likewise, can be no principle of being, but only a principle of knowledge." Accordingly, he later insists on the "intrinsic nonobjectivity" of the I, such that there is no sense in which the I is a thing that exists outside of its own awareness or intuiting of itself.[7]

In the "Introduction" we find a hint concerning how natural science is to be conceived as passing over into "pure intelligence" when he notes that although the "concept *of nature* does not entail that there should also be an intelligence that is aware of it," nevertheless, it is a "necessary tendency of all *natural science* . . . to move from nature to intelligence."[8] We must understand this claim against the background of Schelling's insistence not to limit natural science to physics (a limitation commonly attributed to Spinoza) but to include sciences such as chemistry and especially the then-rapidly developing biological and medical sciences. What Schelling seems to be suggesting is that once a mode of naturalistic interpretation of the world is adopted, all phenomena, and ultimately the phenomena of human intelligence itself, will be brought under its purview. And the consequences of such comprehensiveness of natural explanation can be seen in his later comments on materialist approaches to knowledge.

Materialism aims to explain cognitive and other "spiritual" phenomena in terms of matter. It starts, then, from the presuppositions that "primordial being can transform itself into knowledge (*ein Wissen*)" and must show that "even representation [*Vorstellung*] [is] itself a kind of being." But this, he says, is simply to turn matter "into a phantom [*Gespenst*], the mere modification of an intelligence, whose common functions are thought and matter. Materialism itself thereby reverts to the intelligent as that which is primary" and so "no longer differs . . . from transcendental idealism."[9] The idea seems to be that having adopted the materialist orientation to our cognition, any refusal to acknowledge the status of our representations *as* a "modification of intelligence," that is, as knowledge, and to regard it as *merely* a "kind of being" would bring down the whole presupposed edifice of scientific knowledge by undermining its meaning *as* knowledge. The only alternative here for Schelling seems to be some teleological account of the emergence of the mind *from* matter. Thus, "the completed theory of nature would be that whereby the whole of nature was resolved into an intelligence," which would reveal nature's highest goal of becoming "wholly an object to herself" as well as allowing us to recognize that "nature is identical from the first with what we recognize in ourselves as the intelligent and the conscious."[10] In this way, the adoption of a consistent naturalistic cognitive science must end up with the investigator occupying that "idealist" position from which the system started, that position from which self-consciousness was regarded as primordial, self-positing, and inexplicable. While a year later Schelling was to appeal explicitly to Spinoza's geometric method, here his only explicit reference to Spinoza's philosophy is to describe it as the only consistent form of dogmatism and a form of thought that can "endure only as a *science of na-*

ture [*Naturwissenschaft*], whose last outcome is once more the principle of transcendental philosophy."[11]

I will return to such a teleological understanding of nature, but first let us look at the converse movement that leads from the viewpoint of idealism to the adoption of its opposite, that of realism. The idealist approach must start from the principle of self-consciousness, but as we have seen, this cannot be conceived as knowledge of the objective existence of some kind of thing or being (the Cartesian cogito, for example) existing independently of that knowledge of which it is the principle and in which it is immanent. Here Schelling seems to align himself with those aspects of Kant's first critique, which are antiphenomenalistic, even directly realistic, in their orientation. Just as Kant's transcendental idealism implied a type of "realism," an "empirical" one, so too does Schelling's Fichtean idealist starting point, the infinitely "self-positing *Ich*," make sense only as the principle immanent within knowledge of an *objective* natural world. However, it turns out that for Schelling any actual subject's epistemic activities (what he calls its "producing" [*produzieren*]) presupposes its own "real," that is, practical acts. This is so because cognitive experience, as Kant had insisted, relies on a subject's self-reflexivity, that is, its grasp of itself as the unitary subject *of* that experience. Schelling took this to mean that not only had the subject to be able to *intuit* itself, it had to intuit itself *as* intuiting. But this can not be done *directly:* "The self cannot simultaneously intuit and intuit itself as intuiting" (54). It can only be done by a subject intuiting itself in some intuitable (that is, perceivable) products of its own *real* activity—real activity because only such acts have intuitable products. But real activity has an "ideal" aspect, in that real acts are themselves consciously and intentionally performed and are directed into a world as known. Thus, "*both activities,* the real and the ideal, *mutually presuppose each other*" (40–41). Moreover, it is on a basis of such mutually presupposed but opposed aspects of the self that the similarly mutually presupposed but opposed idealist and realist perspectives on the world are founded: "Just as the two activities reciprocally presuppose each other, so also do *idealism* and *realism*." Hence Schelling seems to have incorporated Fichte's opposed idealist and dogmatist philosophical positions back into Fichte's own transcendental idealism by somehow equating the idealist perspective with the theoretical component, and realism or dogmatism with the practical component. "Theoretical philosophy is therefore idealism, practical philosophy realism, and only the two together constitute the complete system of *transcendental* idealism" (41).

Schelling's approach to the task of making explicit the implicit "realist" dimension of Fichte's transcendental idealism is clearly inspired by Kant's own proffered solution to the unification of theoretical and practical philosophy in the third of his critiques, the *Critique of Judgment,* of 1790. For most of its course, Schelling's *System* generally retraces Fichte's steps in the 1794 *Foundations of the Entire Science of Knowledge.* Thus, first of all, in his Part 1 Schelling gives his version of the deduction of the fundamental principles from Part 1 of the *Foundations.* After a short "General Deduction of Transcendental Idealism" (Part 2), there unfolds the "System of Theoretical Philosophy" (Part 3) followed by a "System of Practical Philosophy," generally paralleling the structure of Fichte's 1784 text, "Part 2: Foundation of Theoretical Knowledge" and "Part 3: Foundation of Knowledge of the Practical." But then Schelling's text goes on to two further short sections touching on natural teleology and then art (still "according to the principles of transcendental idealism"), that is, essentially the topics (but in reverse order) of the two parts of Kant's third critique, the "Critique of Aesthetic Judgment" and the "Critique of Teleological Judgment." In general, it is Kant's idea of the peculiarity of "reflective" nature of both aesthetic and biological judgment, types of judgment in which the mind recognizes something of its own unity and purpose in a represented external object, that allows Schelling to find the space for a variety of biologically conceived "dogmatism" that can capture Fichte's implicit, but unsayable, realism.

Organic Nature in Schelling's *System* of 1800

What Schelling seems to have been seeking in philosophy of nature is a form of realism not prey to the self-conception that Fichte saw as characterizing "dogmatism." He seems to have been seeking a form of realism in which one was capable of conceiving of objective or "real" existence as the presupposed basis of the self's free theoretical and practical activities, as a realm offering a foothold to the I's activities rather than as antithetically opposed to them. Hence the attraction of Kant's conception of the *organic* (as opposed to mechanistic) order of "natural purposes." Such an idea of a realm that allowed the recognition of nonconscious purpose in its products would allow the conception of an objective living process "whereby the objective world is produced [as] at bottom identical with that which expresses itself in volition, and *vice versa.*" (11–12). The underlying identity of the real and the ideal must therefore

> display itself [*sich darstellen*] in the former's [i.e., nature's] products, and these will have to appear as products of an activity at once *conscious and nonconscious.*
>
> Nature, both as a whole, and in its individual products, will have to appear as a work both consciously engendered, and yet simultaneously a

product of the blindest mechanism; *nature is purposive, without being purposively explicable.*—The philosophy of *natural purposes*, or teleology, is thus our point of union between theoretical and practical philosophy. (12)

In appealing to a conception of the biological world based on the irreducibility of the notion of natural purpose, Schelling was appealing to an emerging and productive paradigm within German biology of his time, a paradigm identified by Timothy Lenoir as the "teleo-mechanist program" of the Göttingen school.[12] This approach, which Lenoir describes as emerging with the work of the naturalist Johann Friedrich Blumenbach, had adopted as a heuristic or regulative principle the Kantian idea of natural purpose as developed in the *Critique of Judgment*. Effectively this meant that *methodologically* one accepted the notion of functional organization as primitive and irreducible.[13] For the organic world, organization had to be regarded as a primary given and as understandable on the analogy with human artefacts since it could not be further explainable in terms of mechanical causation.[14] For Kant, this was because of the linear nature of mechanical causality, a linearity that could not be adapted to the complex patterns of circular and reciprocal interdependencies encountered in the organic realm. Presupposing the notion of organization, however, one could then search for its distinct patterns and underlying schemata across organisms, arrange such patterns into homological series, trace its transformations and developments within developing organisms, and so on. Such an approach, sometimes referred to as "transcendental morphology" or "transcendental anatomy," allowed the growth of the disciplines such as embryology and comparative anatomy, giving them a conceptual space that, according to Lenoir, avoided the extremes of reductionism and vitalism:

> Incapable of forming the notion of a purposive organization from the principles of a causal-mechanical explanation, reason must take organization as its starting point in understanding the mechanical interaction of the parts. It is because nerve substance is first experienced as an organized body with particular functions that an investigation of the electro-chemical mechanisms at the basis of its functional organization can take place, not vice versa.[15]

The teleomechanist program was not vitalistic, because, while permitting talk of *"Lebenskräfte," "Bildungstrieben"* and so forth, such "life" and "formative" forces were not thought of as forces beyond or added to those underlying physiochemical forces constituting them; rather, they were conceived as existing at the level of specific organizations of such underlying forces.[16]

But Schelling's conception of natural teleology nevertheless differs from that of Kant. For Kant, the peculiarity about the reflective judgments that grasp nature in terms of such organization is that the mind conceives of those purposes immanent to organization on the analogy with its *own* purposefulness.[17] In this heuristic move nature is regarded "as if" it were purposeful: I am able to recognize purposes in nature only because I myself have purposes and express them in action. Natural purpose, therefore, must be conceived as in some way "projected" onto nature: "We only borrow this causality from ourselves and attribute it to other beings without wishing to assume that they and we are of the same kind."[18] But this cannot be how things stand for Schelling. Once the ideal and real sides of the subject-object and the corresponding theoretical and practical perspectives are seen as *mutually presupposing,* the analogy cannot work in such a one-sided way. That one-sided theoretical view of nature in which the mind unifies it as an object of knowledge at the expense of finding itself within nature, that is, the Fichtean stance that grasps the objective world purely as lifeless mechanism, is itself a one-sided abstraction from the more basic world experience of an embodied subject (a "subject-object"), which necessarily combines both ideal and real perspectives. We may, like Kant, think of purpose as "borrowed" from ourselves and then projected onto or added to the world with the adoption of the biological stance. However, the mind itself has to be embodied in a nature of "recognized natural purposes" if it in turn is to *have* purposes, that is, the purposes that could serve as the *vehicle* for the original analogy. Hence it would seem that the purpose cannot simply be "projected" onto nature, it must also be *found* there.[19]

In the first decades of the nineteenth century Schelling came to enjoy a significant influence within certain German medico-biological circles. While once it was common to dismiss the idea of "Schellingian nature philosophy" as having any positive effects there, there has been a tendency to reverse this diagnosis. Thus, for example, Clarke and Jacyna in their history of the development of modern neuroscientific thought have argued that Schellingian ideas were positively implicated in the most significant theoretical shift in neuroscientific conceptualization in the first third of the nineteenth century when the traditional Galenic "top-down" conception of neural structure and function became replaced with a "bottom-up" approach, essentially setting the broad parameters for research up unto the present.[20]

While proponents of the traditional view had regarded the brain as the seat of reason and as controlling the body by way of the peripheral nerves and the spinal cord, supporters of the bottom-up view essentially reversed this, coming to see the brain more as a complex arrangement of the same functional units (ganglia) that operated relatively automatically at "lower" levels of the system, such as within the spinal cord, or even outside the central nervous system and within the "vegetative" or "autonomic" nervous system.[21] We have seen one important example of this phenomenon in the early history of the "cerebral reflex" idea, where it had been the more nature-philosophically oriented German researchers, or their English followers like Carpenter and Laycock, who had accepted that the model of the spinal reflex arc could be applied to cortical activity and the "higher" operations of thought. In contrast, opponents such as the English physician Marshall Hall had preserved a more Cartesian distinction between the operations of thought and the mechanistically understood bodily reflex.[22]

One suggestive way in which the idea of the mutual reciprocity of the somatic and the mental is expressed in *System of Transcendental Idealism* concerns the way Schelling applies the Kantian biological idea of "organization" from the "real" (biological) realm back to the "ideal" cognitive aspect of the subject-object by talking of cognition itself as a function of the "organization" of the flow of mental representations (*Vorstellungen*).[23] Somewhat in the way of the twentieth-century "cybernetic" approaches to cognition, Schelling conceives the organization of a succession here as the function of its *constraint:* thus cognitive organization is "succession confined within limits and presented as fixed," "succession hampered [*gehemmte*] and . . . coagulated [*erstarrte*]." Furthermore, the succession is constrained in such a way that some underlying form or configuration recurs: organization is thus *self-reverting* succession. This self-reverting character of the succession of states characteristic of organization, mirroring the self-reversion typifying the process of intellectual intuition, also lifts it from *linear* causal processes: organization is thus "excluded . . . from mechanism" and subsists "not merely as cause or effect, but through itself, since it is at once both cause and effect of itself."[24] In this way the elements of mental life originating in sensation can be raised to a "higher power" because of the role they play within the organization of some higher form of cognition.

Consistent with his general approach to cognition Schelling's starting point here is again Kant. Cognition must involve more than the mere having of ideas or even their "association"; rather the stream of *Vorstellungen* must be subject to some sort of constraint, for Kant, the imposition of a unifying rule or concept on the sequence of intuitions encountered in inner life. Also, like Kant, Schelling is concerned with how such a constraint is required for an "objective" content to be contained within such a flow of *Vorstellungen*, how the organization of the flow can somehow reflect the world itself.[25]

inal limitation, without being aware of this intuition, or the latter itself again becoming an object for the self. In this phase the self is entirely rooted upon the sensed, and, as it were, lost therein" (55, 61).

Without extensive relations, such a content would not be "grasped" in time or space at all. Indeed, "grasping" itself implies a separation of self and content that is not present here. Given the fact of the self's being "rooted upon" or "lost in" sensation, such sensation cannot strictly be considered as a *Vorstellung* at all as there is nothing that "stands before" the self. As purely intensive, sensation cannot be that state that provides a certain foundation for knowledge as it is regarded by empiricism. For any epistemic relation to be instituted, self and content must separate for the self. In Schelling's formula, both self and object must "revert" upon themselves to form separable, and so relatable, terms. What empiricism does not explain is "the reversion of the self *upon itself*, or how the latter relates the external impression to itself as self, or intuitant. The object never reverts into itself, and relates no impression to itself; for that very reason it is without sensation" (62).

From a philosophical point of view, that is, from within the Fichtean deduction, we can see that the content is a product of the self's two activities, but this becomes apparent for the self itself only in a more complex form of experience, "productive intuition." Productive intuition is a genuinely cognitive form of experience, a form that essentially corresponds to perceptual knowledge with its implicit separation within consciousness between knowing subject and known object. For Fichtean reasons, here the self must become self-conscious of itself *as* sensing (must "revert" to itself), the precondition for any content to become "objective" and grasped representationally: "The original sensation, in which the self was merely the sensed, is transformed into an intuition, in which the self for the first time becomes for itself that which senses, but for that very reason ceases to be the sensed" (69). In contrast to sensation, in which self was absorbed in the content, here self and content are grasped as separated self-contained things "in themselves." Those dogmatic realist philosophers who explain the mind's epistemic contents in terms of the effects on the mind of external objects are themselves presupposing or thinking from the position of this particular form of experience, while from the Fichtean position the object that constitutes the real from this point of view is itself a posit of the ideal side of the real-ideal subject.

With the deduction of a separable object or content of cognition, Schelling goes on to "deduce" further from the dynamics of the subject with its two opposing forces, the structure of the content formed within this structure of experience. Essentially this can be thought of as the

structure of a purely *extensive* "matter," an extensivity that is the negation of the pure *intensivity* of sensation. The first aspects of matter so deduced are in Kantian term, those two pure intuitions presupposed by
any experiential content, space and time.[31] But here it must be remembered that this deduction proceeds from a reflection on such content
from the philosophical perspective; it is not yet part of how any real subject limited to this type of productive intuition will understand itself or
its object.

As seen, the "subject" (if we can call it that) of "original sensation" does
not grasp itself in time or space at all: It simply feels itself as that purely
intensive sensation into which it is absorbed. This is the Schellingian
equivalent of the mental life of Freud's infant prior to the onset of the "reality principle." In fact for Schelling, original sensation is clearly not considered a form of experience for any conceivable consciousness, for we
conceive of forms of consciousness in terms of the general nature of their
contents and original sensation has no separable "content." Nevertheless,
it is a model of immediate self-intuition, implicit in all forms of consciousness, although never a *component* of those forms of consciousness. As
James was to say, "No one ever had a simple sensation by itself."[32] It is
productive intuition that provides the first actual form of consciousness in
Schelling's "history," and within productive intuition, sensation, in a
somewhat paradoxical sense, "appears" as a form of experience that has
been lost. Schelling describes the subject of this form of experience as
feeling itself "driven back" to that "earlier" now unexperienceable sensation, "driven back to a stage of which it cannot be conscious. It *feels* itself
driven back, for it cannot in fact go back." Such a "feeling of being thus
driven back to a stage that it cannot in reality return to is the feeling of the
present. . . . This feeling is no other than that which we describe as self-
awareness. All consciousness begins with it, and by it the self first posits
itself over against the object."[33]

With this Schelling is clearly trying to get at that difficult phenomenological point about the nature of self-awareness within the temporal flow
of consciousness about which James agonized first with his idea of the
"stream of consciousness" and then with that of the "vertical" axis of the
two dimensions of experience. For Schelling there is a deep connection between the temporality of self-awareness and the ability to represent or
posit external objects. "If the self is to recognize the boundary between itself and the object as contingent," he notes, "it must recognize this as conditioned by something that lies wholly outside the present phase" (102).
That is, it must recognize this boundary as conditioned by some unrecov-
erable sensation no longer present as a component *within* experience.
What we call self-awareness simply *is* this feeling of sensation denied, de-

nied because it is past; it is neither some awareness of the self as a thinking thing, as Descartes thought, nor the experience of some immediate sensory content understandable as a direct state of the self. Kant had thought of time as the medium of inner sense, but Schelling identifies time and the self in a much more intimate way: "In that the self opposes to itself the object, there arises for it the feeling of self-awareness, that is, it becomes an object to itself *as* pure intensity, as activity which can extend itself only in one dimension, but is at present concentrated at a single point; but in fact this unidimensionally extensible activity, when it becomes an object to itself, is time. Time is not something that flows independently of the self; the *self itself* is time conceived of in activity" (103).

Time and the Organism

From within Fichtean transcendental idealism it is clear that we will never reach a point at which that self-intuition of which the self of our investigation is capable will capture both its real and ideal aspects since the self cannot intuit itself as intuiting. But with the conception of self-intuition as a feeling of the present that "as activity . . . can extend itself only in one dimension" Schelling alludes to the way this problem is to be resolved. A self will be capable of intuiting itself as active only once it has acted, that is, acted in the "real" sense of practical action in the world. Schelling will thus ultimately pass from the theoretical to the practical (and so "realist") perspective because it is in the product of the self's real acts that it will find a fixed content that it can identify as its own. Not surprisingly then, the idea of the subject-object's orientation to the "now" will also feature in the context of the subject-object's real activity: The "now" is the temporal point of the subject-object's active intervention in the world, the point from which it can "step out from its own producing and intuit itself as productive" (121).

But this orientation to the present must put a further constraint on the subject-object's experienced relation to time. The subject-object must start out from the "now" but nevertheless be capable, as Kant described, of locating itself within time, and so as viewing itself from without, that is, from a perspective other than that of the now. But just as it is never actually sunk in the now as agent (akin to the absorption of the self in the content of original sensation), neither is it ever capable of achieving a viewpoint that is entirely indifferent to the now, achieving that eternal "god's-eye view" of time as an infinite series, from which it reflectively grasps its locatedness. Schelling's denial that the human cognitive endeavor is ever capable of achieving an aperspectival god's-eye view of the world "in itself" is expressed in terms of his idea that intuition is never capable of "absolute synthesis." As in Fichte, intellectual intuition can never be satisfied, it could only ever function for us as a kind of norm or ideal. But what implications does this notion have for the goal of understanding oneself in one's actions? Schelling treats the issue in the following way. The content of one's real act is something that unfolds within representations within time, but there must be a limitation within this series for the subject-object to grasp its objective act as an intuitable whole; otherwise the act would never gain an intuitable form. What Schelling is alluding to here is the impossibility of seeing the consequences of any action in such a way that those consequences simply formed an infinite causal chain flowing into an indefinite future. One would *never* understand what one's act qua real intervention into the world was if one could never come to a point in which one said that while *this* result of the action can be identified as part of it, some other distant unforeseen result could be dismissed. Action must be conceived as having an *end*. Thus the series of representations giving the meaning of one's act must be, Schelling notes, "an infinite succession within this limitation" (121); it must be a succession that "only appears to proceed in a straight line, and constantly flows back into itself" (122). It is in the time of a teleologically conceived *organic* nature that this type of cyclical, but infinite, movement is to be found. Such an appeal to organic nature here is based on an idea much more substantial than any ancient idea that nature moves in cycles; rather, it reflects more a Kantian teleology of the organism, an idea close to the modern conception of homeostasis. Let us focus on the parallelism that Schelling is trying to develop between his deduction of "productive intuition" and the inverse "nature-philosophical" account of the forms of experience attributable to an "evolutionary" series of organisms as revealed in the biology of his time.

"Evolution" of the Organism

At this point in his deduction Schelling has reached the idea that the proper place of intelligence must be as embodied in specifically *living* bits of the world. This conception enables him to line up his "deduced" history of self-consciousness with a parallel developmental series of living things as found in those pre-Darwinian "evolutionary" comparative schemata of the late eighteenth and early nineteenth centuries. Here humans are located at the extreme end, or top, of a more or less continuous series stretching from the simplest to the most complex forms of life. Early in this sequence are plants, which exhibit a cyclical mode of temporal existence. It is only within the lowest orders of the animal kingdom, however, that we start to encounter a receptivity to the immediately present in the "feeling sense" of touch (*Gefühlsinn*), the crude tactile and light sensitivity attributed to simple organisms such as worms within such taxonomies, a receptivity to the present which has implications for the organism's relation to time.

Schelling here is careful to insist on the nonrepresentational character of this "feeling sense." It does not testify to "a power of acquiring representations through impressions from without, but merely to their relationship with the universe, which may be broader or more confined" (124). There is strictly no representing of anything involved in such organisms because within *Gefühlsinn* there can be no requisite separating of self and object allowing something to stand objectively before the animal. That is, the lowest animals seem to instantiate that "original sensation" implicit in all consciousness but not present as any component of such states. The feeling sense does not *represent* anything for the animal, but nevertheless, we can still think of the universe as in some sense exhibited or *presented* (*dargestellt*) within animals of the lowest order. In the sensitivity or reactivity of these animals to their environment, such animals exhibit the universe condensed to aspects of their immediate environment, although they do not exhibit it *to* themselves, that is, do not *represent* it. But from here "if we move upwards in the scale of organization, we find that the senses gradually develop in that order in which, by means of them, the world of the organizations is enlarged" (124). Schelling here refers to the work of Carl Friedrich Kielmeyer, one of the initiators of the teleomechanist paradigm, who had attempted to classify the animal world on the basis of comparative anatomy and physiology, in particular in terms of how the internal structure of different kinds of organisms expressed a different configuration among the basic forming forces (*Bildungskräfte*) of reproductive power, irritability, and sensibility.[34] Schelling mentions how different sense organs take in lesser or greater ranges of the environment (hearing having a smaller range that the "godlike sense" of vision, for example). Clearly, the idea of increasing complexity of the real organization of the organism is meant to include the increasing integration between the sensitivity of the different senses as the number and differentiation of senses increases. But sight defines the limit of the organism's world. As we will see, what is needed for the cognitive scope of the animal to transcend this boundary is the capacity for judgment, the capacity definitive of the human form of understanding.

This Kielmeyer-derived series thus provides Schelling with forms of life within which the deduced forms of experience can be located, and the types of intuition that we can retrospectively find in nature can now be schematized into a simple triadic structure, typical of Schelling's "potence" structures:[35] First, is sensation with its object, the simple undifferentiated "stuff" or "matter" (*Stoff*) (that is, internally undifferentiated, and undifferentiated from the self); next is productive intuition, with its object (*Materie*); and the third stage "is characterized by organization." With this third stage Schelling does not separate out form of intuition and its object but refers simply to "organization," noting that the third state is merely the second (*productive intuition*) raised to the "second power [*Potenz*]."[36] That is, it is productive intuition that has become intuited as such by the organism. Seemingly, while organization necessarily characterizes the *process* of which the *Anschauung* of matter in productive intuition is the product, in the third state it is demonstrated (to the philosopher, not the object intelligence under examination) how such a form of experience is necessarily the outcome of its underlying organizing forces. The type of self-intuition that is implicitly or potentially present here would be one that intuited the self *as* self-organizing. This type of intuition would be explicitly developed in the third epoch of self-consciousness, which moves from reflection to "absolute act of will," clearly, the distinctively human epoch.

We might pause here to reflect on the implication of Schelling's schema for the mind–body relationship as conceived by subsequent nineteenth-century advocates of psychophysical parallelism such as Fechner or Mach. In Fechner, the elements of cognitive states are thought of in terms of the notion of "sensation." It is precisely the *intensive* qualities of sensation that suggested, of course, that they could be correlated with the degree of stimulation of the nervous system. But for Schelling, although the inner life of some basic organisms may be so conceived, for the Kantian reasons with which we are now familiar, the content of human conscious mental life cannot be thought of as constructed from sensation in this way. In fact, not even that of the higher animals capable of their own perceptual representation of the environment could be so conceived. What this important differentiation of merely sensing organisms from actual intuiting (perceptually representing) ones, in which sensation is elevated to the second power, seems to signify is that the cognitive *functioning* of those higher organisms must be understood in terms of some higher level of "organization," than that which consists of the mere succession of sensation states. What is needed is precisely a type of organization existing at the level of that organism's *representations*. And Schelling does have such a notion. With his ideas of mental content (starting with sensation) as capable of being raised to a higher "power" in which it becomes representation, and with his general idea of the nature of "organization" itself, he seems to anticipate the modern notion of the separation of an organism's "informational" states from its most basically energetic ones.

The Schellingian Unconscious

We have seen some of the similarities that Schelling's thought bears to Freud's various attempts to chart psychical reality with his notion of the unconscious. But Schelling, too, explicitly uses this concept, and it is from this quest for philosophical systematicity incorporating theoretical and practical philosophy that his use of the idea of the unconscious arises. Theoretical and practical forms of consciousness appear to presuppose irreconcilable assumptions: from the theoretical standpoint we assume that "there is a passage from the real world into the world of representation," whereas from the practical standpoint we assume that "there is a passage from the world of representation into the real world." That is, theoretically we consider our representations as determined by objects or states of affairs, practically, those objects or states of affairs as determined by our free representations. The resolution of this contradiction, which he describes as the *"highest* task of transcendental philosophy," is to be sought only in the higher discipline to which theoretical (idealist) and practical (realist) philosophical perspectives are a part. Again, it is Schelling's somewhat Spinozist belief in the "parallelism" of nature and intellect that is crucial here. The answer to this highest task must lie in the assumption of a "predetermined harmony" between ideal and real, a harmony in that the "activity, whereby the objective world is produced, is at bottom identical with that which expresses itself in volition and *vice versa.*" Willing is *conscious* productive activity, but its worldly parallel is, of course, "without consciousness."[37]

This concept looks like Spinoza's idea of mind and nature as two aspects of one underlying substance, but such an idea of "substance" is, of course, a "dogmatic" assumption in the eyes of a transcendental idealist, and so Schelling takes this supposition a step further so as to bring its insights back within the orbit of transcendental philosophy. If such a predetermined harmony is the case, then the identity of the real production with the ideal production will

> display itself in the former's [nature's] products, and these will have to appear as products of an activity at once *conscious and nonconscious.*

> Nature, both as a whole, and in its individual products, will have to ap-
> pear as a work both consciously engendered, and yet simultaneously a
> product of the blindest mechanism; *nature is purposive, without being purpo-*
> *sively explicable.*—The philosophy of *natural purposes,* or teleology, is thus
> our point of union between theoretical and practical philosophy. (12)

A philosophy of natural purpose per se would seem to be a Spinozist
type of nature philosophy, but with the idea of nature as "purposive,
without being purposively explicable" Schelling has set himself up to return to the framework of transcendental philosophy and so to close the
circle with nature-philosophy via Kant's aesthetics:

> But now the system of knowledge can only be regarded as complete if it re-
> verts back into its own principle.—Thus the transcendental philosophy
> would be completed only if it could demonstrate this *identity*—the highest
> solution of its whole problem—*in its own principle* (namely the self).
>
>   It is therefore postulated that this simultaneously conscious and noncon-
> scious activity will be exhibited [*aufgezeigt*] in the subjective, *in conscious-*
> *ness itself.* (12)

Fichte had established the duality of conscious and unconscious with the
idea of intellectual intuition as that which while being a condition of all
consciousness was not *itself* conscious.[38] Schelling too stresses this inter-
linking of the conscious and unconscious throughout his discussion of
mental life. Thus: "To come to consciousness, and to be limited, are one
and the same. Only that which is limited me-ward, so to speak, comes to
consciousness: the limiting activity falls outside all consciousness, just because it is the cause of all limitation."[39]

As seen, the idea of unconscious mentality had been at least implicit in
Kant. Schelling not only makes this notion explicit but also clearly prefig-
ures some of Freud's central concepts such as those of primal repression
('*Urverdrängung*') and "splitting" (*Ichspaltung*).[40] For Freud "repression
proper" involves the self's turning away from painful representations, the
forcing of them into the "unconscious." But this process relies on the existence *of* an unconscious with contents capable of "attracting" these once
conscious representations. The presupposed process in which these original nuclei came to inhabit the unconscious is primal repression, while
"splitting" is essentially the same process seen from the side of the psychical apparatus itself in which it becomes separated into its conscious
and unconscious parts (or, in the second topography, into the agencies of
the id, the ego, and the superego).

We can see how for Schelling this is all part of the "movement" from the nonrepresentational state of "original sensation" to the representational consciousness of "intuition." With the development of representational consciousness, the "self as sensed," states Schelling, "would . . . be expelled [or repressed, *verdrungen*] from consciousness, and something else opposed to it would take its place."[41] That which takes its place is, of course, intuition, which is a representation of something other to the self.[42] But to be an "*Ich*" requires self-intuition, thus the human organism is thereby driven though a variety of further cognitive forms in an attempt to secure self-intuition.

Along with such a process of internal splitting of the subject into conscious and unconscious aspects, Schelling's picture provides for structural differences between conscious and unconscious thought. Thus beneath the more "rational" structures of consciousness can be divined mental operations based on the polar opposition of the "limited" and "unlimited" psychological forces. Originating in his studies in chemistry in the 1790s, Schelling had come to see the principle of polar opposition as implicit in all natural and spiritual phenomena, the "principle of polarity" becoming one of the defining characteristics of all nature-philosophical explanation. For Schelling, mythopoeic thought in particular was conceived as structured by analogically mappable polar oppositions, such as day/night, waking/sleepfulness, male/female, life/death, and so on. Indeed, the very distinction between conscious and unconscious thought itself could be accounted for in terms of this universal polarity, as well as the analogical connections of the unconscious to madness, dreams, sleep, and so forth.[43] In turn, the operations of the unconscious could be seen as grounded in the functioning of the "vegetative" or autonomic nervous system, producing the underlying schema for romantic psychiatry stretching from Schubert to Jung. In the post-Freudian period, similar ideas, although shorn of their nature-philosophical framework, are seen in Lacan's notion of the unconscious as structured by the "metaphoric/metonymic" structures of language, and Matte-Blanco's ideas of the "bipolar logic" of the unconscious.[44]

Within recent philosophical interpretations of Freud, such an account of the unconscious as having its own distinct structure of thought has been subjected to a type of neo-Kantian criticism by Marcia Cavell: Without the capacity to structure an objective (that is, propositional) representational content, why should such structures be regarded as structures of *thought?*[45] But considered in the Fichtean way, such structures might be seen as among the preconditions that *allow* thought to achieve representational objectivity. That is, they may reflect the Fichtean demand that a thinker be in some sense in immediate nonrepresentational contact with his or her own self considered *as* a thinker. Schellingian nature-philosophy eventually fell into disrepute because of the way such mythopoeic thought structures articulating intellectual intuition were taken as symbolic vehicles of some deep metaphysical truths. But in Fichte, the immediate self-relation of intellectual intuition was always in tension with the reflective, representational structures of a consciousness of objective contents, leading him to reject the direction taken by Schelling. Moreover, this constitutive tension within consciousness was restored within Hegel's philosophy, with Schelling's polarities coming to be constrained in turn by the structures of reflective, propositional thought.

7

Hegel, Affect, and Cognition

One might expect that of all the influential nineteenth-century ideal-ists whose ideas might contribute to an understanding of the mind in the late twentieth century, Hegel, the "absolute idealist," would be last on the list. Nevertheless, in late-twentieth-century discussions about the nature of mind some directions that have surprisingly overt Hegelian aspects emerged. Thus, philosophers of cognitive science such as Daniel Dennett and Andy Clark have conceived of thought processes as going on "outside" the head, while starting from the philosophy of language, Hilary Putnam and Tyler Burge have conceived of thought *contents* as, in an analogous way, external to the minds of individual thinkers.

Reflecting on the similarity of our brains to those of the members of other, related species, Dennett has posed the question of the source of our disproportionately greater intelligence. His answer is to stress our

> habit of *off-loading* as much as possible of our cognitive tasks into the envi-ronment itself—extruding our minds (that is, our mental projects and activ-ities) into the surrounding world, where a host of peripheral devices we construct can store, process, and re-represent our meanings, streamlining, enhancing, and protecting the processes of transformation that *are* our thinking. This widespread practice of off-loading releases us from the limi-tations of our animal brains.[1]

Similarly, Clark sees distinctively human thinking in terms of processes that are "scaffolded" upon complex physical and social structures outside the brain. This means that "much of what we commonly identify as our mental capacities may . . . turn out to be properties of the wider, environmentally extended systems of which human brains are just one (important) part."[2]

For both Dennett and Clark it is our spoken and written words that make up the most significant of these scaffolding structures, and their approaches to this issue of the interaction of brains and environment are crucial to their criticisms of the older "classical" cognitivist picture in which cognitive process are conceived on an analogy to serially operating, explicitly rule-governed symbol processing of single computers. Drawing on more recent "connectionist" models of computation, they see this classical view as making the mistake of assuming that the processes observed in linguistic interaction must mirror and be derived from analogous processes within the brains of the language users themselves.[3] Such a view does not take sufficient note of the actual ways in which brains in interaction with independently existing external structures are responsible for the complex expressions of which we are capable. Thus, for example, the fact that in language we find propositional representations that are minimally reliant on context and modality[4] should not lead us to automatically assume that similar representations are actually encoded in the brain—a view cohering with Dennett's denial of the psychological reality of "propositional attitudes."[5]

Starting primarily from considerations to do with the semantic properties of words, Putnam and Burge have reached somewhat analogous conclusions about mental contents.[6] In their "externalist" approach to mental content, they have argued against the view that meanings and intentional kinds (that is, kinds of propositional attitudes) can be individuated psychologically. For them, the content of a concept or of a proposition cannot be specified in ways that make no reference to actual worldly objects and events, that is, things external to the brain. We might see such externalist positions as rehearsing, in late twentieth-century terms, ideas expressed in others ways in James's "direct realism"; but beyond this, along with the work of Dennett and Clark, the externalist position exhibits broad similarities to ideas about the nature of mind put forward almost two centuries ago by Hegel in his insistence on the dependence of individual "subjective" mind on the historically accumulated and culturally transmitted structures and processes of the "objective mind." And for Hegel, as for Dennett, Clark, Putnam, and Burge it was language that was the most important of these mind-bearing cultural scaffolds.

Hegel's Spirit

It might be considered that in his *Anthropology* Kant had set the general parameters for subsequent idealist thought about how a finite organism could become a subject capable of representing a world. Fundamentally, the interpretation of sensational states as components of representational states required a transformation of the organism's felt self-relation into some type of self-conception: a representing subject had to grasp itself as that abstract "transcendental ego" capable of unifying its mental contents into a coherent representation of an external world. By invoking the necessity of having one's self-relation linguistically mediated by the symbol "I," Kant was implicitly appealing to communicative sociality as a condition of subjecthood.

In Fichte, as seen, all cognitive activity is grounded in that feeling which is the immediate subjective presentation of the embodied self's "checked striving," but in the context of his theory of rights in *Foundations of the Science of Right* in 1796, he too invoked a more communicative framework. Asking what it is that can limit a subject's action such that it respects the rights of others, Fichte answered with the notion that the subject must be "determined to determine itself" and expanded on this apparent paradox with the idea that the subject must somehow freely limit itself in response to a *requirement* or a *demand* that is "addressed" to it *as* a free agent. It will be qua addressor of such a demand that some other agent will be recognized and responded to as the bearer of rights.[7] But if, as has been suggested by some interpreters, this demand is seen as identical with the original *Anstoss* or "check," we can see that much more than the recognition of rights of others is at stake. The recognition of the *address* of others will be contextualized in the original act of self-positing and will take the form of some interpretative positing to the other in response to this address.

With this move Fichte might be seen as an originator of the type of hermeneutic conception of cognitive and practical subjectivity found in a thinker such as Hans-Georg Gadamer, for whom belonging to essentially *dialogical* relations constitutes critical conditions of human subjectivity. But this conception, as I have argued elsewhere, is also central to Hegel's notion of *Geist* or spirit.[8] Human organisms can be freely knowing and acting subjects not only by virtue of any natural determinations, but also because of their belonging to distinct patterns of normative intersubjective relations in which they recognize each other as intentional beings and recognize themselves *as* recognized by others. The linguistic medium of this interaction is signaled by Hegel's famous description of spirit in the

Phenomenology of Spirit as "this absolute substance which is the unity of the different independent self-consciousnesses which, in their opposition, enjoy perfect freedom and independence: 'I' that is 'We' and 'We' that is 'I'." The self can only recognize itself as an "apperceptive I" through the mediation of another, another whom it recognizes as such a being and who recognizes it in the same way: "Self-consciousness exists in and for itself when, and by the fact that, it so exists for another; that is, it exists only in being acknowledged."[9]

For Hegel, such reciprocal recognition stands at the heart of those various forms of human interaction whose objectifications make up the scaffolding of "objective spirit." In the *Phenomenology of Spirit* the project that started with Fichte's "pragmatic history of the mind" and Schelling's "history of self-consciousness" is developed into a massive history of the accumulation and transformations of such mind-supporting cultural scaffolding, that spiritual "second nature," into which any individual "subjective" spirit is born. Hegel's increasing distance from Schelling and his rejection of the broader "romantic" recourse to the immediacy of "feeling" in contrast to the mediated "determinacy" of conceptual thought and reflection have often been taken as involving a blanket rejection of the role of "feeling," "subjectivity," and "immediacy" in thought. However, Hegel clearly continued to work within that broad framework first initiated by Fichte in which mental content was anchored in a realism wherein feelings could exist only in living knowers open to the world despite the fact that such feelings could play a determinate role in psychological life only when taken up into a subject's conceptualized positings.

We see the particular role played by feeling in Fichte's theory of recognitive spirit in his treatment of the institution of the family in *Elements of the Philosophy of Right*. Here, the fundamental social bond holding individuals within the family is love, "spirit's *feeling* of its own unity."[10] As the vehicle for the reproduction of the species, the family is close to being a "natural" or merely biological grouping, but it is in truth a fundamentally "spiritual" structure, an instantiation of the " 'I' that is 'We' " and " 'We' that is 'I'." For that reason Hegel is not satisfied with the modern romantic conception of love in which naturalistically conceived feeling is regarded as its essence; in criticism of those who would see the public, ceremonial nature of marriage as something external and inessential, he regards the explicit linguistic avowals and symbolism of the marriage ceremony as that which brings love to its mature, institutionalized form.[11] This type of valorization of the explicit and conceptualized in the life of the mind has commonly led to Hegel being condemned as a hyper-rationalist or "panlogicist" for whom all feeling must be replaced by a conceptually determinate form of mentality. However, such criticism misses the mark: Hegel was rather concerned that a wedge not be driven between feeling and concept in mental life such that feeling would thereby become sequestered in an inaccessibly private subjective realm. In fact, as Daniel Berthold-Bond has pointed out, for Hegel this was the condition that was at the core of madness.[12] Thus Hegel describes as a form of psychopathology (mania or frenzy) a particular disorder of "self-feeling" (*Selbstgefühl*) in which the individual is "aware of the disruption of his consciousness into two mutually contradictory modes"—the immediate self-relation of feeling on the one hand, and the universal abstract conceptuality of the "I" on the other—but is "unable to find himself in the *present* from which he feels himself repelled and, at the same time, bound."[13] More like Fichte, Hegel seems to have taken feeling as constituting a demand for an interpretative, conceptualizing response. Feeling is not replaced by concepts in such responses to the world; rather, feeling underlies the very activity of conceptual positing. Fichte's phrase, "Intuition *sees*, but is *empty*; feeling *relates to reality*, but is *blind*,"[14] might equally be said of Hegel's attitude here as well.

With these considerations in mind we might appreciate the resonances of Hegel's comments in his *Philosophy of Spirit* on a postulated science of "psychical physiology" that would study the "system by which the internal sensation comes to give itself specific bodily forms."

> The most interesting side of a psychical physiology would lie in studying . . . the bodily form adopted by certain mental modifications, especially the passions or emotions [*als Affekte geben*]. . . . In physiology the viscera and the organs are treated merely as parts subservient to the animal organism; but they form at the same time a system for the somaticization of spirit [*ein System der Verleiblichung des Geistigen*], and in this way they get quite another interpretation.[15]

It is in terms of this somaticization that feeling could come to function within the system of communicative recognition of the other. The smile, for example in response to another's presence, is not simply the somatic side of an inner feeling of happiness, but also an outward sign within which the other can recognize him- or herself as the "addressee" of that smile.[16] Without an external, public, side feelings could come to have no significance for it is only then that they can be taken up within the communicative relations between individuals and thereby become articulated with the concepts of their linguistic positings. Only then could they come to function in human forms of life *as* feelings.

Hegel's concept of recognition has elements of Kant's idea that the mind can recognize itself in nature in teleological judgments, but as in

Schelling's transformation of Kant, such recognition of the worldly *expressions* of mentality can no longer be seen as secondary to the mind's own immediate understanding of its own contents, its intentions, for example. The recognition of others *as* agents, which allows the reflective recognition of oneself *in* their expressions, thus becomes a *condition* of mentality per se. This seems to be how pieces of "nature" are incorporated into existing structures of spirit. An infant will be drawn into the world of human communication by the fact that its parents typically interpret and respond to its expressions, its smiles, for example, as a form of intentional expression—in the case of the smile, as a form of love directed at themselves. That is, Hegel argues for the ineliminability of what Daniel Dennett has called "the intentional stance";[17] and for Hegel as for Dennett, this stance cannot be grounded on the assumption that such a recognized being is *already* an intentional creature independently of that recognition. But while this seems to mean for Dennett that the mind is not "really" intentional, for Hegel it rather implies that without recognition *there is no mind*. That is, the adoption of this stance not only allows the mind to recognize more mind as nascent in a part of living nature, but also helps to create the conditions within which that nature can become yet another bit of mind.

Thus by being drawn into the game of recognition and the adoption of the intentional stance the infant can start to extend its own self-awareness from mere self-feeling to ultimately the reflective "I." To grasp oneself as an "I" is to grasp oneself as a participant within a process of communication in which one's own and others' mental contents are grasped in terms of their shareable (that is, "modality-neutral") propositional structure. But for Hegel, as for Jackson and Freud, this process cannot be one in which earlier forms of self-awareness are simply left behind and replaced by the later. The human being came into the world as an embodied, feeling individual organism, and it is in terms of this spatio-temporally specific singular organic thing that it remains attached to its purely natural origins.

We might gain a perspective onto Hegel's overall orientation to feeling and its relation to conceptuality by glancing at a well-known passage from his later lectures on aesthetics in which he assesses the role of feeling in aesthetic judgment. There he notes that any type of aesthetic theory that sees art simply as intended to arouse pleasant feelings does not lead very far because:

> feeling is the indefinite dull region of the spirit; what is felt remains enveloped in the form of the most abstract individual subjectivity, and therefore differences between feelings are also completely abstract, not differ-

ences in the thing itself [*die Sache Selbst*]. . . . Feeling as such is an entirely empty form of subjective affection.[18]

This disqualification of feeling on the basis of its individual and subjective characteristics might at first sound like that of Kant, for whom subjective affections do not at all enter into the representation of an object, but merely express a subjective state in a noncognitive way. But for Hegel the problem with the feeling-based approach to art has to do with the way that it "is satisfied with observing subjective emotional reaction in its particular character."[19] This popular approach reflects upon feeling as merely a state of the subject and so abstracts feelings from the cognitive role they play in experience. Hegel's attitude to bare feeling here is thus somewhat akin to Kant's attitude to unschematized sensation. Feelings are relevant only when they are integrated into the cognition of some worldly thing—here an artwork. We might appeal to James's formulation and say that for Hegel feelings are relevant only when the subject can regard the "local changes and determinations" of his or her own body as an integral part of his or her embodied knowing self. It is thereby that these somatic states "pass for spiritual happenings."[20] That is, Hegel's is a fundamentally Fichtean attitude to feeling that treats feeling as an occasion for the conceptualized determining of the object deemed responsible for it. Hegel's criticism of the romantics' valorization of feeling was thus not a criticism of that unconscious feeling that Fichte had dealt with in terms of "intellectual intuition"; rather, it was a criticism of its exclusion and isolation from other components of psychological life.

Hegel, Tomkins, and Psychoanalysis

Hegel saw himself as standing at the endpoint of the development of post-Kantian idealism (and much more besides). Regardless of how we evaluate this self-assessment we do find in Hegel a development and systematization of those suggestive ideas from Kant, Fichte, and Schelling linking together feeling, conceptualization, and intersubjective recognition and communication into a rich and powerful model of the mind. As a consequence Hegel has been described by the psychoanalyst Arnold Modell as the "first intersubjective or relational psychologist."[21]

Among the post-Freudian psychoanalysts it was probably Jacques Lacan who was most heavily indebted to Hegel's recognitive theory. In Lacan's account, the illusory traps of an imagistic form of self-relation (the "mirror phase") can be avoided only by the individual's passage into a form of experience and self-conception thoroughly mediated by language proper.[22] However, Lacan's theory has been criticized for its overreliance on the linguistic and cognitive dimensions of mental life (conceived, moreover, in a highly idiosyncratic way) and its relative silence on affect.[23] But we might find a greater approximation to Hegel's views on the relations of affect to cognition in the work of Silvan Tomkins, a psychoanalytically influenced American theorist of affect who held out against the tide of high cognitivism in the 1960s, 1970s, and 1980s. Significantly, however, Tomkins was not a theorist who resisted the more general "cybernetic" revolution that had been one of the founding disciplines behind the development of the "mind's new science." Yet in contrast to those for whom the operations of the mind were conceived as indifferent to the material nature of the system in which they were instantiated, Tomkins insisted that it not be forgotten that a "living system such as a human being is a feedback system rather than a communication system, and therefore the freedom of such a feedback system must be distinguished from the formal theory of the information . . . of a communication system as measured by Shannon."[24]

Tomkins follows James's view that it is somatic feedback to the brain that is constitutive of affective states. Although the "inherently acceptable or inherently unacceptable . . . essentially aesthetic characteristics" of affective responses are "in one sense, no further reducible,"[25] such phenomenal characteristics are nevertheless identified with somatic states: "My original observations of the intensity of infantile affect, of how an infant was, for example, seized by his or her own crying, left no doubt in my mind that what the face was doing with its muscles, and blood vessels, as well as with its accompanying vocalization, was at the heart of the matter. This seemed to me not an 'expression' of anything else but rather the major phenomenon."[26]

Again as in James's account, the ultimate explanation of the existence of affect is to be found in the evolutionary history of the species, which accounts for the existence of "basic affects," such as the positive affects of "interest-excitement" and "enjoyment-joy," the negative affects of "distress-anguish," "fear-terror," "shame-humiliation," "contempt-disgust," and "anger-rage," and what he called the "resetting" affects of "surprise-startle." Such basic affects are primarily described in terms of their bodily expression—the smile of joy, for example, or the cry of distress—with the phenomenological characteristics of each of these resulting from the feedback from changes taking place in bodily viscera, muscles, and skin, and in particular the muscles and skin of the face. But such a locus of affective responses means that affect will be "at once individual and private and social and shared nonverbal communication."[27] Therefore, these basic reflex responses fit immediately into characteristic communicative patterns between individuals. Affect stands at the interface of the most private and public, facing both ways.

It is thus that Tomkins can conceive of affect, in a somewhat Hegelian way, as bridging the biological and the cultural by virtue of its underpinning structures of recognition and communication. While biologically grounded, affect is freed from the determinacy of the biological: by means of the feedback that the feeling subject receives in the affective responses of others as well as its own further response to this feedback, a child comes to have affective responses to its own first-order affects (as when, for example, one feels shame at one's fear). It is by such mechanisms of mediation that the child is induced into existing patterns of socially codified "ideo-affective postures."[28]

Freud's postulation of the ubiquity of the expression of the sex drive was, of course, for him consequent to the adoption of the naturalistic perspective of the evolutionary history of the human species. For Tomkins, however, evolutionary explanation applies equally well to the affect system uncoupled from the specificity of the drives:

> Humans are among those animals whose individual survival and group reproduction rest heavily on social responsiveness, and the mutual enjoyment of each other's presence is one of the most important ways in which social interaction is rewarded and perpetuated.
>
> The smiling response and the enjoyment of its feedback along with the feedback of concurrent autonomic and hypothalamic responses make possible a kind of human social responsiveness that is relatively free of drive satisfaction, of body site specificity of stimulation, and of specific motor responses other than that of the smile itself.[29]

Whereas Freud saw love primarily as sublimated sexual drive, Tomkins's account of love is far more Hegelian in that it is dealt with in terms of the mutual reinforcement and the mutual recognition of affect:

> Smiling crates a *felicité à deux* similar to and also different from that created by the enjoyment of sexual intercourse. In sexual intercourse, the behavior of each is a sufficient condition for the pleasure of each individual for himself and at the same time for the pleasure of the other. This dyadic interaction is inherently social inasmuch as the satisfaction of the self is at the same time the satisfaction of the other.
>
> In the smiling response, as we see it first between the mother and her child, there is a similar mutuality, except that it is on the affect level rather than through mutual drive satisfaction, and it operates at a distance rather than requiring body contact. . . . Later, when the child's development is suf-

ficiently advanced, both parties to this mutual enjoyment are further rewarded by the awareness that this enjoyment is shared enjoyment. This is
mediated through the eyes. Through interocular interaction both parties become aware of each other's enjoyment and of the very fact of communion
and mutuality. Indeed, one of the prime ways in later life that the adult will
recapture this type of communion is when he smiles at another person, who
smiles back at him and at the same time the eyes of each are arrested in a
stare at the eyes of the other.[30]

But the communicative interaction even between parent and child cannot
be *all* like this, as the parent–child interaction must also become the place
within which the socialization of the child's affective life into culturally sanc-
tioned ideo-affective postures takes place. Thus the parent must, to a greater
or lesser extent, take hold of the "conjoint opportunity and obligation to
mold the child to some norm." In this, "the parent sets himself in opposition
to the child and bestows upon the child the sense that positive satisfaction is
necessarily an epiphenomenon, consequent to effort, to struggle, to renunci-
ation of his own immediate wishes. His own feelings and wishes are de-
valued in favor of some kind of behavior which is demanded of him."[31]

Such a process is, of course, analogous to what Freud described as the
imposition of the "reality principle," the acquisition of the "secondary
process" of conceptual thought. Essentially, the child must come to understand its feelings in a new way. Within the affect system alone, as within
Freud's "primary process," there can be no distinction between the affect
itself and anything other than it that it signifies. Positive affect *is* the de-
sired state, and its achievement is its own success. But such feeling must
come to be grasped, in Tomkins's terms, as itself "epiphenomenal," that is,
as not itself counting for what is ultimately real but rather as indicating
some other more objectively defined achievement or satisfaction. Put another way, the child must come to understand its own affective states as
representational, as indicative of some satisfaction that can be viewed from
an external, third-person point of view. In so doing, it will come to thereby
incorporate its own affective states into the conceptual representations of
language. And as a part of this achievement, it must learn to move, as Kant
had pointed out, from an immediate *feeling* of itself, to a concept of itself as
an "I" who is the bearer of such shareable representations.

Affect, Reason, and the Polarities of the Primary Process

Such were the peculiar characteristics of the affect system that allowed
Tomkins to contest the cognitivist approach to affects dominating Amer-

ican psychology in the 1960s and 1970s, the so called appraisal theory. Cognitive theory agreed with common sense (or, as Tomkins puts it, the views of "Everyman" dominant for the last two thousand years) in its grounding of affect in rational *appraisals*.

> Everyone knows that we are happy when (and presumably because) things are going well and that we are unhappy when things do not go well. When someone who "should" be happy is unhappy or suicides, Everyman is either puzzled or thinks that perhaps there was a hidden reason, or failing that, insanity. There are today a majority of theorists who postulate an evaluating, appraising homunculus (or at the least, an appraising process) that scrutinizes the world and declares it as an appropriate candidate for good or bad feelings. Once information has been so validated, it is ready to activate a specific affect. Such theorists, like Everyman, cannot imagine feeling without an adequate "reason."[32]

But it is often the case with affects that their objects *cannot* be understood as providing "good reasons" for the having of those affects. "There is literally no kind of object which has not historically been linked to one or another of the affects. Positive affect has been invested in pain and every kind of human misery, and negative affect has been experienced as a consequence of pleasure and every kind of triumph of the human spirit."[33] For Tomkins, affect is generated by processes essentially equivalent to the Freudian *primary process*, a process not held in the thralls of the reality principle, but nevertheless presupposed by any application of it. As seen, with his notion of the primary process Freud was essentially recovering a tradition of thought characteristic of the romantic psychiatry from a hundred years earlier, with its appeal to the symbolic language of dream and myth. In fact, the core idea behind Schelling's notion of this symbolic language of the "night side" of life can be traced back to Kant.

In his aesthetics, while denying that aesthetic experience conveyed any "truth," Kant had nevertheless held that aesthetic qualities could play a "symbolic" role in the "exhibition" of concepts that allowed of no intuitive content. Thus in the final section of the first, "aesthetic" half of the *Critique of Judgment*, in a discussion of how beauty can symbolize morality, Kant discusses the various ways in which concepts can be made sensible (their "sensibilization"—*Versinnlichung*). Empirical concepts can be exemplified by intuitable particulars, in the way that an actual cat, for instance, can exemplify the concept "cat," while in the case of pure concepts of the understanding, concepts can be *schematized*. For *rational concepts* or "ideas," however, "absolutely no intuition can be given that would be adequate to them" to allow their theoretical cognition.[34] Nevertheless, ideas can be *indirectly* sensibilized by the use of analogy. Kant calls this latter type of sensibilization "hypotyposis" or *Darstellung* (presentation or exhibition). While the direct exhibition of concepts is *schematic,* the indirect, analogical exhibition of ideas he names "symbolic."

Kant notes how God can be sensibilized in this way, but comments how, at a more mundane level, the German language is replete with such analogical exhibitions. Words such as *Grund* (foundation), *Abhängen* (to depend upon, to be held from above), *woraus Fliessen* (to flow from, used in the sense of to follow), and *Substanz* (substance, as used by Locke as the bearer of accidents) all express concepts "not by means of a direct intuition but only according to an analogy with one." Kant even allows that this mode of representation (*Vorstellungsart*) can count as *cognition,* but only for *practical,* not theoretical, reason.[35] But given the post-Kantian desire to unify theoretical and practical reason, we can see how the notion of an analogically based *Darstellung* of the unknowable was likely to be relished.

Indeed, symbolic *Darstellung* was to become central to Schelling's philosophy. As seen, in his *System of Transcendental Idealism* Schelling had defined the philosophy of art as "the universal organon of philosophy—and the keystone of its entire arch"[36] because it is only in aesthetic production that conscious and unconscious forms of productive activity are unified. This aestheticization of thought becomes fully explicit in his *Philosophy of Art* (1802–1803), in which both art and philosophy are seen as standing on the same level as parallel forms of *Darstellung* in which the absolute is exhibited.[37] With this identification Schelling seems to draw all the consequences that are commonly considered as serious grounds for concern with such an aestheticized view of philosophy. Linking the ideas of *Darstellung* and of "intellectual intuition," he openly talks of the philosopher as a type of seer, an individual "whose intellectual intuition should be directed only toward that particular truth that is concealed to sensual eyes, unattainable and accessible only to the spirit itself." The parallelism between philosophy and art has given a new meaning to "intellectual intuition": while art allows the objective intuition of ideas in its works "philosophy intuits these ideas as they are *in themselves.*"[38] Furthermore, this aestheticization is accompanied by an apparent valorization of the symbolic and analogical forms of mythopoeic thought over more standardly conceptual and argumentative forms of reason.

While Schelling's flight into the artistic and mythological *Darstellung* of the absolute can seem to reflect a complete abandonment of that "realistic" dimension of his thinking reflected in his philosophy of nature for a type of mystical platonism, we should not lose sight of the continuing connection here to his more realistic thought about the natural world.

Schelling's focus on art, symbolism, and myth is bound up with a concern with finding the objective, material forms within which those mental processes dealt with as embodied within the organism could gain an extra-organismic or cultural form. "Just as reason becomes immediately objective only through the organism, and the eternal ideas of reason become objective in nature as souls of organic bodies, so also does philosophy become objective through art. . . . For just this reason, art is to the ideal world what the organism is to the real world."[39]

It was in this sense that the typically "romantic" stress on affectively laden analogical thought and myth could combine with the new "bottom-up" view of the nervous system in which the "higher" cortical activities could be thought of as grounded in and continuous with the unconscious activities of the "lower" subcortical centers. Such a unified view was what had recommended Schelling to a generation of medical and psychiatric thinkers who would apply such ideas to psychopathology and other human phenomena.

As seen, Freud, following Jackson, understood waking experience as structured by an internalized language. Propositional speech mediated by associations established at the level of the cortex became in some way superimposed on more primitive reflexes working at the level of subcortical ganglia, thereby "re-representing" the sensory-motor relations "represented" at those subcortical levels. For Schelling, too, language had a crucial role in the mind's powers. Providing an "external body" for thought it opened up the thought of an individual to the perspectives of others, thus allowing thinking to push beyond the "original limitation" provided by the singularity of the organism. Thus while from our embodied perspective we have a particular view onto the world, "everything resides as one in language, regardless of the perspective from which one views it." But language is still rooted in nature: "Viewed from the one side, language is the direct expression of something *ideal*—knowledge, thought, feeling, will, and so on—in something *real*, and is to that extent a work of art. Yet viewed from the other side it is just as definitely a work of nature, since it is the one necessary form of art that cannot be conceived as being invented or generated by art. Hence, it is a natural work of art, just as more or less everything produced by nature is."[40] As a work of nature it rested on the polarities of a somatically based language of the unconscious that manifested itself in dreams and mythologies.

In this century, anthropological interest in the polarities of mythopoeic thought has continued, stripped of its Schellingian framework. For example, early in the century the Durkheimian Robert Hertz, described the formal structures implicit in the mythological thought of preliterate communities in such ways.[41] For Hertz, the fundamental evaluative polarity upon which binary pairs of "primitive classification" were set up was that of "right" and "left." In Hertz's account, such asymmetry was then projected onto those binary categories such as "male"/"female" and "sacred"/"profane" regarded as in some way homologous to "right" and "left." While acknowledging some truth in the claims of those who grounded this latter distinction naturalistically on the process of cerebral lateralization, Hertz nevertheless postulated a fundamentally cultural determination for the valorization of the right side over the left, and for that of the corresponding terms of those homologous polar opposi- tions.[42]

Later, in the 1950s and 1960s, the idea a "pensée sauvage" constructed from an armory of culturally given binary oppositions capable of struc- tural transformations was made popular by the work of Claude Lévi- Strauss. Drawing upon the then-recent information-theoretic approach, such "magical" or "symbolic" thought, he claimed, constituted a mode of thought "parallel" to that of modern science and provided for premodern communities a taxonomy operating at the level of sensible properties. For Lévi-Strauss, the elements of such thought were signs which "lie half-way between percepts and concepts."[43] In analyses of shamanistic cures in non-Western cultures as well as the phenomenon of "voodoo death," he stressed the link between symbolic thought and somatically constituted strong states of emotion.[44] But while Lévi-Strauss's "structural" analyses have been criticized by Pierre Bourdieu and others for overintellectual- izing such practices, the general principle of a distinct "logic" of such mythopoeic thought based on the primacy of polarly contrasted sensible properties has received continuing support among a number of anthro- pologists.[45] Furthermore, it can be recognized in the history of philosophy as well. As G. E. R. Lloyd has pointed out, the basic structures of pre- Socratic cosmology are those of polarity and analogy: The contrasts hot/cold, dry/wet, light/dark, unlimited/limited, and so on provided a type of intuitional-conceptual scaffolding for judgments about the world that, in projecting such terms in analogical ways, give a thinkable articula- tion to the world's sensory appearance.[46] And as Bourdieu has pointed out, while largely disappearing from the (at least, explicit) structure of our cognitive discourse, such polarities have nevertheless retained a central place in the *aesthetic* discourse of the West.[47]

In Hegel's work it is in the context of his exposition of the "categories of Being" within the *Science of Logic* that we find his most explicit account of this type of polar thought. Paralleling both Fichte's and Schelling's at- tempts to find a basis for all thought in intellectual intuition, Hegel starts his logical presentation (his logical *Darstellung*) with a reflection on a type of "thought determination" that is meant to capture its content in an immediate and nonrepresentational way—"being."[48] Hegel's notorious arguments here revolve around the instability of the conceptual distinction between "being" and its opposite, "nothing." "Being," he argues, always passes over, or always has already passed over, into nothing. Most commonly Hegel's claims here have been treated as exemplifying a precritical metaphysical stance that aims to establish something about "being" (that is, that it is really "nothing," whatever that could mean) on the basis of the concept alone. But in the context of post-Kantian idealism it can be seen as making the point that it is only in the act of positing some "other" that the type of immediate self-awareness that Fichte described as intellectual intuition can have any determinate form. But, of course, "being" has no "other": the all-inclusive scope of this category means that nothing can stand opposed to it. It might *appear* to contrast with some other category called "nothing," but actually contrasts with nothing at all. (We might say that the concept of "being" by taking *any* empirical content gains none in particular.) This is not to say that "being" and "nothing" are meaningless categories, but rather, that these categories cannot be understood in isolation from others needed for making "determinate" any thought content.

In his further categorial "deduction" it is only when Hegel comes to the category of *Dasein*, "determinate being," that we find concepts capable of having a limiting other. Particular concepts belonging to this category pick out particular "somethings" via some experienceable qualities: "Determinate Being [*Dasein*] is Being with a character or mode—which simply *is;* and such unmediated character is quality."[49] Hegel counts *demonstratives* like "this" in terms of this category, and what allows a "this" to achieve a determining limit is the fact that it can be contrasted with a "that." The important thing about this pair of quasi-concepts, and what separates them from the pair "being"/"nothing," is that with them both members of the contrast can be *exemplified:* we typically use demonstratives such as "this" and "that" as a type of pointing to empirical presences. It is precisely such opposing sensuous exemplars that allow the type of analogical projection seen in mythopoeic thought.

With this form of sensuous opposition Hegel is on Fichtean territory. In the *Wissenschaftslehre* Fichte, struggling with the circularity of the problem of how feelings can be the foundations of empirical judgments, inquired into the nature of the *ground* of a distinction between, say, something *sweet* and its *bitter* opposite. The distinction seems to inhere within the qualitative characters of the *feelings* involved, but feelings are not *as such* determinate. One aspect of the answer that Fichte had offered there involved the idea that our feeling-prompted positings are *necessarily dual*, the concepts involved related as polar opposites. Thus we can only posit *sweet* things if we can also posit *bitter* ones that limit and thereby determine the feelings involved in the first positing. Something sweet and something bitter are "both something, but each is something different; and only thereby do we first arrive at posing and answering the question, *what* are they? Without counterposition (*Gegensetzung*), the entire not-self is something, but not a determinate or particular something, and the question, *what* is this or that? has in such a case no meaning whatever; for only through counterposition does it obtain an answer."[50] We might say that without this structure of counterposition, the "entire not-self" is *mere* being—merely *is*. With something so indeterminate, the question "*what* is it?" (like the question "What is being?") has no answer.

And so for Fichte, feelings must in some sense come in opposing pairs as "without reflection on *both* there could be reflection on *neither of them*, as feelings."[51] Hegel's logic of being, effectively starting from the categories of determinate being, essentially reflects this necessary "counterpositing" on the basis of such opposed underlying feelings or sensations. Significantly, it is this characteristic of the conceptual pairs of the logic of being, the fact that *each* member can be equally exemplified in a "this" or "that" able to be posited as the immediate cause of this or that felt sensation that separates these categorial pairs from those of the second book of the *Logic*, "The Doctrine of Essence."

Like those of Being-logic, the concepts of Essence-logic also come in contrastive pairs; but these pairs, such as in the distinction reality/appearance, exhibit a structure in which only *one* member of the pair is directly exemplifiable in something, the something held as responsible for the felt sensation involved. The second, nonappearing, member of the pair is *conceptually* rather than perceptually grasped—comprehended rather than apprehended—as something hidden "behind" or "beneath" the first, somehow responsible for it, but not, in terms of experiential access, on the same level as it.[52] We find in "reflective judgment" the type of thought that exemplifies these categories. Thus we do not immediately "apprehend" the curative or poisonous properties of a plant, for example, as we would its *color*; rather, we *comprehend* it in terms of the perceivable effects that that plant has on other things. That is, the structures of Essence-logic are those structures required for thought to achieve the "objective" (we would now say "propositional") content of a Kantian judgment. Any implicit "indexical" reference back to the conditions of perception have here been transcended. And one important aspect of this transition from the structures of being to those of essence that is connected with this move from an apprehended to a comprehended content, is a decrease or elision of the *phenomenality* of the mental states involved. Thus in terms of immediate perceptual apprehension, while there is something that it is like for a rose to be red, there is no equivalent "what it is like" for a plant to be curative or poisonous.

Hegel's overall position in the *Logic* seems to be that the thinking of creatures like ourselves, finite and embodied and yet capable of reason and freedom, requires *both* types of categorial structure; each is necessary but not sufficient for our cognitive lives. Thus at the end of "Being-logic" we find that the type of thought dealt with there in fact presupposes the "mediated" structures of essence, while at the end of essence we find the converse—its structures rely on some sense of immediacy that its categories are unable to supply. But given Hegel's background in the thought of Fichte, this is what we would expect. The type of "immediacy" of one's cognitive relation to oneself requires the mediated positings of representational consciousness, while they, to *be* representations, presuppose a nonrepresentational cognitive form, that form in which we are aware of states of ourselves. Neither structure without the other can produce consciousness or thought.

But considered in isolation from the structures of essence, those of Being-logic are peculiar indeed. The needed symmetry between its conceptual pairs would seem to imply that the "somethings" thought with the concepts of this logic could never be conceived objectively as any such something is determined *by* its (subjectively dependent) experienceable quality: it "is what it is in virtue of its quality, and losing its quality it ceases to be what it is."[53] That is, Being-logic simply does not have the conceptual resources for grasping a thing's identity other than by linking it to some characteristic quality it exhibits to us, and so loss of its identifying quality must be interpreted as the annihilation of that something, or a change in a thing's quality must mean that it has, like the alchemist's element, transmuted into something else.[54] This is Hegel's equivalent to the logic of myth and dreams, a "logic" that, as it were, takes all metaphors literally. But unlike Schelling and Jung, Hegel does not assign to it any priority over reflective thought. More as in Freud, it calls forth for conceptual articulation in the "secondary processes" of essence—we might say that for Hegel, "where 'being' is, there 'essence' shall be." But also as in Freud, the structures of being are not simply replaced by those of essence. As negated or superseded and preserved (*aufgehoben*) within reflective thought, they transform the latter into a form of thought that gives a place to both, a form of thought dealt with in the third book of the *Logic*, the "doctrine of the concept."[55]

It is with this last claim, to have achieved a truly "scientific" form of thought that supersedes all empirical science, that Hegel has been subjected to the fiercest of criticism by an army of subsequent philosophers of diverse persuasions. Several recent interpreters of Hegel have, however, seen in this idea more a continuation of Kant's critical philosophy than the hubristic metaphysics of which Hegel is usually accused.[56] Regardless of this overarching issue, however, remains the question of what might be learnt from that aspect of Hegel's thought regarding the difficult and pressing issues of the relation of affect and cognition in the life of the mind.

8

The Relevance of Idealist Psychology
in a Darwinian World

For much philosophy of mind in this century, nineteenth-century idealist thought has represented little more than an example of radical
wrong-headedness—as a fundamentally antiscientific nostalgia for an
earlier world in which humans could be consoled by the idea of their
occupying a determinate place in a divine order. Materialism, surely,
has been on the side of science, and idealism its opponent. However,
nineteenth-century thought about the nature of mind reveals a paradox of
the degree to which progressive scientific thought was indebted not only
to idealism but also to the nature-philosophy that sprang from it. If we accept the views of cognitive theorists such as Kitcher and Brook about the
advanced nature of Kant's "cognitive science," it is not difficult to see
why. Against the background of the then-rapidly developing biological
sciences and further that of the development of the cultural sciences, post-
Kantian idealists such as Fichte, Schelling, and Hegel grappled with a
problem familiar to current cognitive scientists—how to reconcile a "top-
down" task-performance view of cognition with a view of human beings
as elements of a culturally shaped biological world. In this earlier post-
Kantian world the conception of cognition as somehow grounded in the
type of immediate self-feeling that a human individual had because of his
or her biological makeup was the lynchpin holding together the Kantian
demand for self-consciousness on the one hand and the biological nature
of human thinkers on the other. And it gave a very determinate shape to
the conception of the relation of feeling to cognition.

While we commonly regard both James and Freud as belonging to the advance guard of the thought of the twentieth century, I have suggested that it is instructive to see both as late products of the nineteenth. Both struggled with a problem that subsequently disappeared from most psychological and philosophical discourse but that has been again revived. This was the problem of articulating two different dimensions of human conscious mentality, the phenomenal on the one hand, and the intentional and cognitive on the other. During the nineteenth century, a common way of understanding the nature of phenomenality, made popular by writers such as Lewes, Spencer, and Fechner, had involved the quasi-Spinozist move of grasping the realm of consciousness, now understood as sensation, as paralleling from an inner perspective what from an outer perspective was a neurophysiological reality. But this conception was at odds with a concurrent post-Kantian approach to the mind in which sensation per se was rendered unconscious, to reappear in a different, functional role within representational states of consciousness. In struggling with these issues, both James and Freud retrieved various aspects of the complex philosophical psychologies bound up in earlier idealist philosophy, putting them to work in new ways, and setting them in the context of newly discovered facts. And thereby both made great advances in the understanding of the human mind.

Should we see this idealist participation in psychological knowledge simply as just an historical curiosity, however? A phenomenon with no real relevance to *present* psychological and philosophical discussions of the mind? This, I contend, would surely be unwise. Fundamental problems about the relation of phenomenality and cognition have resurfaced, particularly in studies of that region of mental life where phenomenality is at a premium, affect. After decades of domination, the strong cognitive paradigm has been challenged by Jamesian somatic-feedback theorists. Moreover, within theories of cognition itself, more biologically appropriate cognitive "architectures," such as that of connectionism, have challenged those earlier ones conceived more on the model of the serial computer. Along with this, more and more workers in the field have come to question the reigning orthodoxy of the brain as a mechanism for performing "computations" over "representations." We might say that if Kant was at the philosophical advance guard of a "top-down" cognitive science, surely his successors were at that of a "bottom-up" response that faced the question of how to get the Kantian representational mind back into the organism.

Yet it remains undeniable that James and Freud do stand on our side of the huge break separating twentieth-century thought from early-nineteenth-century idealism initiated by Darwin's revival of evolutionary theory under the revolutionary form of a theory of natural selection. Not only did James and Freud reject idealist thought, but also both explicitly appealed to Darwinian underpinnings for their conceptions of the mind. Of course, the same applies to a recent figure like Silvan Tomkins. Did not the publication of *The Origin of Species, The Descent of Man,* and, especially for our purposes, *The Expression of the Emotions in Man and Animals,* erect a barrier to what could be possibly retrieved from an earlier period when divine purpose was understood as pervading the world?

Idealism and the Evolution of Human Language and Thought

The question of the relations that could possibly exist between post-Kantian idealism and evolutionary-based naturalism is far too complex to be more than touched upon here. Nevertheless, I believe that it is clear that any simple antithetical opposition of evolutionary naturalism to post-Kantian idealism is misplaced. For one thing, Kant's "idealism" was not that of the earlier *metaphysical* opposition between views that asserted either "matter" or "ideas" to be what ultimately existed. As seen, Fichte saw his position as a "real-idealism or ideal-realism," asserting a complementarity between realist and idealist perspectives that was continued by both the early Schelling and Hegel. But irrespective of these broader questions, there exists the issue of the degree of continuity at a more concrete level between Darwinian explanations of the organism and the sorts of idealist and nature-philosophical ones that preceded Darwin. Commenting on the Darwinization of neuroscience, Clarke and Jacyna have asserted that in the case of the evolutionary claim of continuity of the human brain with those of lower vertebrates "all of these principles were established long before the publication of Darwin's theory and were, in most cases, independent of any theory of physical descent. Their true roots lay in the romantic philosophy of biology we have considered. What can be said is that the Darwinian theory eventually offered a posteriori justification for the theory of unity of type that later morphologists found useful in justifying their commitment to this dogma."[1] Perhaps this might also be said about the fundamental outlines of much contemporary evolutionary thinking about some of the most distinctive features that make us human: language and propositional thought.

Taking a communicative tack, the post-Kantians from Fichte to Hegel increasingly stressed the role of social existence and its mediation by language in filling out the substance of the normative dimension of human cognition. That evolution-influenced theorists, such as Hughlings Jackson, for example, would see the development of the capacity for language as crucial for understanding the distinctly human mind is hardly surprising. Nor is it surprising that such an orientation among evolutionary thinkers has continued into the present. Among recent accounts of the evolutionary origin of language, a particularly comprehensive one that bears on the relation of affect to cognition is to be found in Terrence Deacon's work *The Symbolic Species*.[2] Bringing a mass of evidence to bear from diverse disciplines such as linguistics, cognitive science, the neurosciences, comparative anatomy, and primatology, Deacon has offered a theory of the evolution of human cognition that, while thoroughly naturalistic, is, in its central claims, striking in its broad resemblance to the nineteenth-century idealist approach.

For Deacon, the most plausible evolutionary story involves the development within our pre-hominid australopithecine ancestors of a simplified symbolic protolanguage that was superimposed upon a qualitatively distinct existing communicative system that they shared with other primates.[3] Deacon takes pains to stress the structural and functional *discontinuities* between these systems, however. While some primates, for example, display innate alarm calls that are specific to distinct predators (eagles, leopards, and snakes, for example, in the case of vervet monkeys), such calls should not be thought of as some type of primitive *naming* of those beings.[4] Utilizing Charles Sanders Peirce's threefold semiotic categorization of "icon," "index," and "symbol," Deacon posits that innate alarm calls function as "indices" of the predators to which they refer, tokens produced in more or less direct spatiotemporal contiguity with their perceived referents. In contrast, human words that function as names are "symbols." Peircean symbols are not related to their referents in such direct ways as are indices, but rather via the mediation of *other* symbols. This in turn means that the symbolic nature of linguistic reference is necessarily related to the compositional or "syntactic" nature of linguistic form.[5]

In Deacon's account, the first appearance of the capacity for this type of symbolic reference among some ancient ancestors demanded new "learning paradigms" and "mnemonic strategies" cognitively distinct from those subserving the simpler indexical form of reference. Neurally, the conditioned reflex is the type of structure that supports indexical reference and that underlies the specific types of learning process ("stimulus generalization" and "learning set" transfer) associated with this type of reference; but quite different forms of learning strategies (involving "logical" or "categorial" association) and, in turn, different forms of neural processing are required for *symbolic* association.[6] However, the acquisition of these symbolic associations presupposes existing indexical ones that "are necessary stepping stones to symbolic reference . . . [which] must ultimately be superseded for symbolic reference to work."[7] This supersession is required because the former associations actually interfere with the latter, creating the need for their suppression in favour of others derived from them. Essentially associations between stimuli from signs and objects have to be displaced by associations among signs themselves, such that these latter come to mediate the former associations, allowing them to now serve symbolic rather than indexical reference. In this process: "What one knows in one way gets re-coded in another way. It gets *re-represented*. . . . You might say we know them both from the bottom up, indexically, and from the top down, symbolically."[8]

Deacon's emphasis on the importance of the qualitative distinction between properly symbolic reference and merely indexical reference parallels the approach of linguists, who typically stress the discontinuities between human languages and the communicative systems of other species.[9] Moreover, his stress on the discontinuity between the learning modes associated with each form of reference parallels Chomsky's early criticism of Skinner's behaviorist attempt to understand the acquisition of language skills on the basis of the conditioned reflex. Thus, like Chomsky, he holds that human infants are born genetically preadapted to learn human language. And yet Deacon rejects the "Chomskian" interpretation of this phenomenon in terms of the evolution of an innate, universal grammar or of a "language acquisition device," as advocated by rival theorists like Steven Pinker and Derek Bickerton.[10] Deacon favors a more "distributed" model of linguistic function, such as that employed by the neuroscientist Gerald Edelman.[11] Furthermore, more like Daniel Dennett and Andy Clark, he stresses the *externality* of language as an artefactual scaffolding for the mind. Chomskians "assert that the source of prior support for language acquisition must originate from *inside* the brain, on the unstated assumption that there is no other possible source. But there is another alternative: that the extra support for language learning is vested neither in the brain of the child nor in the brains of parents or teachers, but outside brains, in language itself."[12]

The "externalist" component of Deacon's account is reflected in his thesis of the "co-evolution" of language and of the brain. After the establishment of some primitive, but symbolic, proto-language, the human brain and this proto-language itself were able to co-evolve according to the type of processes suggested by the early American evolutionary theorist James Mark Baldwin. Baldwinian selection superficially looks Lamarckian, but it is not: initially culturally transmitted forms of adaptive behavior are seen as altering the environment in such ways that new selectional pressures are thereby established. Thus the existence of the cultural artefact of the proto-language within the environment of the species could mean that types of neural processing conferring facility for learning symbolic processes could now be selected. On the other hand, language itself was now able to "evolve" under selectional pressures having to do with its learnability: "Languages don't just change, they *evolve*. . . . Languages are under powerful selection pressure to fit children's likely guesses, because children are the vehicle by which a language gets reproduced,"[13] with the conclusion that "languages have adapted to human brains and human brains have adapted to languages."[14]

Deacon describes the results of this complex natural process as nothing short of miraculous. Moreover, the human mind is miraculous in *just* the sort of way that the idealists understood it to be:

> The evolutionary miracle is the human brain. And what makes this extraordinary is not just that a flesh and blood computer is capable of producing a phenomenon as remarkable as a human mind, but that the changes in this organ responsible for this miracle were a direct consequence of the use of words. And I don't mean this in a figurative sense. I mean that the major structural and functional innovations that make human brains capable of unprecedented mental feats evolved in response to the use of something as abstract and virtual as the power of words. Or, to put this miracle in simple terms, I suggest that an idea changed the brain.[15]

Hegelian *Aufhebung* and Semantic Bootstrapping

We might list among the most central of those "unprecedented mental feats" allowed by truly symbolic reference the ability to make and to understand truth claims. In Frege's account of meaning, the indirectness of symbolic reference is reflected in the fact that individual terms link to the world via their roles in propositions with a truth value. Much the same idea is found in Kant, although there the framework of analysis is predominantly "mental" rather than linguistic representation. Kant had insisted on the normativity of cognitive activity, a normativity lost in causal analyses of cognition, such as the "physiological" approach of Locke, where representation was regarded as the output of causally understood perceptual and associative processes. The starting points here were Kant's ideas that "synthesis" was not to be reduced to association and that rule-governed *judgment* was the unit of cognitive activity. But this notion was purchased at the cost of what seemed an unbridgeable separation between norms of the mind and the causal processes of the body. Post-Kantians from Fichte to Hegel increasingly stressed the role of social existence and of its mediation by language in filling out the substance of this normative dimension of human cognition. For Deacon, the crucial move to be understood is that from indexical to symbolic reference. How might the post-Kantian approach deal with such an issue?

Robert Pippin has pointed out how the idea of "right" from Hegel's practical philosophy has almost a prototypical status in Hegel's understanding of normativity.[16] Perhaps we might use this to reflect on the peculiarity of the Peircean symbolic relation and its distinctness from that of the "index" by an analogy from this more concrete sphere, of the distinction *property* and *possession*. As Kant pointed out, while we may understand possession as a type of natural relation, that is, as a type of empirical spatiotemporal contiguity between the possessor and the thing possessed, in contrast, *property* presents itself as "intelligible." In contrast to the situation with possession, I do not need to be in physical proximity to my property for it to be legitimately *mine:* I own my house just as much when I am away from it as when I am in it.[17] Thus like symbolism, the property relation cannot be thought of as based on simple associative relations. Being an idealist, Kant could accept the reality of such a relation, which, from the point of view of an empiricist like Bentham could only be *fictitious*.

In his *Elements of the Philosophy of Right*, Hegel presents an analysis of the logical relation of property to possession. There are three components to the property relation: first, the process of "taking possession"; next, the *use* of the thing possessed; and finally its *alienation* in the contractual exchange.[18] This paradoxical culminating moment shows something unexpected about the nature of property: a thing is only *really* mine to the extent that I can *transfer it to someone else*, something not captured by any idea of mere possession. But of course the cultural institutions of property would not make sense if individuals never related to their property *as* possessions, consuming them, using them, and so on. And thus while for Hegel the intelligible relation, property, was the "truth" of the empirical, possession, within this Hegelian "dialectical" structure, was somehow "preserved" and yet "suspended" or "negated" (*aufgehoben*) in property.[19] Deacon's idea of an indexical relation that, while presupposed by the more complex symbolic one that was derived from it, must nevertheless be "suppressed" or "suspended," looks like the type of idea that Hegel had in mind with his *Aufhebung*. Conversely, we find that Hegel's account of the relation of property to possession relies centrally on the notion of the sign.

Hegel's analysis of the component process of possession-taking has itself a similar threefold structure. First, there is the action of the "physical seizure" of some singular thing, but the immediacy of such direct physical connection between the body and the thing is its shortcoming: "This mode in general is merely subjective, temporary, and extremely limited in scope, as well as by the qualitative nature of the objects."[20] The next, more developed, form of possession taking involves the imposition of *form*, as in the cultivating of land or the domestication of animals, and this breaks the immediacy of the seizure and allows for the physical separation of owner and owned: "When I *give form* to something, its determinate character as mine receives *an independently existing* externality and ceases to be limited to my presence in *this* time and space and to my present knowledge and volition."[21] But the third and most explicit mode of taking possession of a thing is that of *marking* it with a sign.[22]

Hegel notes that a sign is some thing that does not count as what it is, but as what it is meant to signify. "A cockade, for example, signifies citizenship within a state, although the colour has no connection with the nation and represents [or presents, *darstellt*] *not itself* but the nation."[23] In the *Phenomenology of Spirit,* Hegel had characterized the word in a similar way, stressing its self-negating form of being: the word is "*not* a real existence,"[24] it is something whose existence is exhausted in its function to communicate about some referent (that is, something "not itself") to another. The evanescence of the spoken word thus captures the type of thing or word or a sign is: "through this vanishing it *is* a real existence,"[25] that is, it is precisely this that makes it a "real" word and not just a physical sound. John McCumber has perspicuously summed up Hegel's view of linguistic symbolism thus:

> Its unique form of being is thus to be perceived, heard and understood, by others, and this gives language an intrinsically "universalizing" character. For when someone gives utterance to an idea, those who hear it change and appropriate it. . . . Once expressed to others, my message is no longer my own. It exists in their interpretation, and what is effective in that interpretation is the elements of it understood in common by those others. Such interpretation is thus a "universalizing" activity.[26]

A little like property, my words and thoughts can be truly mine only to the extent that they can become another's. And by taking this type of self-negating being as the mark of mind, Hegel united those current marks of the mental, phenomenality and intentionality, into a single processual whole in which each presupposed but opposed the other. At one end of the process of negation are feelings—those immediate states of the self that James characterized as "local changes and determinations" and "visceral perturbations." At the other are those intentional or representational states, James's "spiritual happenings," which are no longer states of an embodied *me* but states of a mental "I" to whom an objective world is opposed and for whom it is presented. Regarded in isolation from the process, these states look like states of pure intentionless phenomenality, pure affect, on the one hand, and pure aphenomenal intentionality, pure thought, on the other. But for Hegel the reality *is* the process from which the extremes can only be understood as abstractions. Kant had already claimed that as states of myself, sensations were subject to "associations" and had opposed *this* process to that of the "synthesis" undergone by sensations qua contents of my "representations." In Hegel, these two processes reappear as stages of the mind's fundamental process of negation, stages structured by the logics of "being" and "essence," respectively. Hegel's absolute idealism is built on the generalization of the process of negation to its existing at the heart not only of the individual mind but also in the "mind" more generally understood in terms of its objective scaffolding—the practices, institutions, and products making up a culturally transmitted and historically accumulative second nature. Can we square any of this with the Darwinian view?[27]

In fact a surprising amount of this Hegelian perspective seems to survive intact in Deacon's evolutionary account based on the radical transformation of life introduced by the Peircean symbol: "the word made flesh."[28] For Deacon, the human capacity for symbolic speech was superimposed on and presupposed those indexical systems of communication we share with other primates. But these earlier systems need to be at the same time somehow "suppressed" within the properly symbolic dimension of language.[29] Moreover, the three different forms of representation (iconic, indexical, and symbolic) in turn structure three different "forms of consciousness":

> Consciousness of iconic representations should differ from consciousness of indexical representations, and this in turn should differ from consciousness of symbolic representations. Moreover, since these modes of representation are not alternatives at the same level but hierarchically and componentially related to one another, this must also be true of these modes of consciousness as well. They form a nested hierarchy, where certain conditions in the lower levels of consciousness are prerequisite to the emergence of consciousness at each higher level.[30]

We find a similar scenario in the account of the mind given by the neuroscientist Gerald Edelman who also sees the evolutionary shift to human language as responsible for a transformation of consciousness from a perceptually based "primary consciousness" shared with other animals to a "higher-order consciousness" capable of *conceptual* (and not just perceptual) categorization and that "frees the individual from the bondage of an immediate time frame or ongoing events occurring in real time."[31] Edelman refers to the process though which perceptual mental contents are reinterpreted or "re-represented" within higher-order consciousness as "semantic bootstrapping," and again, the picture of mental contents being "semantically bootstrapped" from lower to higher forms of consciousness looks remarkably Hegelian. But are there any reasons to think that any of these lower levels of consciousness are in any way relevant to the idealist understanding of emotion and affect with its peculiarly polar logic?

It seems reasonable to assume that the communication system that was bootstrapped into some early version of our own would have been in many ways like those utilized by present-day primates, those so-called "call-and-display behaviors" that seem to be primarily about the maintenance of specific group relations and that utilize stimuli triggering specific affective responses. What could have been the evolutionary pressures that selected for some form of *symbolic* behavior to add to this earlier system? For Deacon what was crucial was most likely the need to create some sort of mechanism for the establishment of exclusive male reproductive access to mates in complex pre-hominid societies in which those males hunted in groups (the institution of a type of proto-marriage). Thus, at the heart of this emerging human society was a type of symbolically maintained "social contract": "Unlike what is found in the animal world, [marriage] is a symbolic relationship. . . . Marriage, in all its incredible variety, is the regulation of reproductive relationships by symbolic means, and it is essentially universal in human societies."[32]

Deacon's reconstruction is, of course, necessarily very conjectural.[33] But abstracting from its details, we might note his linking of symbolic reference (and thereby, activities like "truth-telling") with the emergence of a distinct type of social relation not found elsewhere in the natural realm, a relation in which members bind their future actions in some form of commitment, hence achieving in practice an analogous form of "freedom from the present" identified by Edelman as being at the heart of "higher-order consciousness." As for Hegel, it is thus by means of symbolic mediation that social life has achieved that peculiar form of normativity that lies at the basis of human forms of thought. Moreover, in his grounding of such symbolism in a context dominated by existing immediate affective communication, we would expect elements of the old system to be bootstrapped (*aufgehoben*) into the new. Thus Deacon notes that in human language it is intonation patterns, accompanying in parallel fashion the more "symbolic" dimensions of speech, that in general subserve the communication of emotion.[34]

Is there any evidence, however, that the affective dimensions of such a bootstrapped system work on the principle of polarity? At least according to Darwin, the answer to this question is "yes." In *The Expression of the Emotions in Man and Animals* Darwin articulated his own theory of the expression of the emotions in man and animals on three "general principles": the principle of serviceable associated habits; the principle of antithesis; and the principle of the "direct action" of the nervous system.[35] In the first, drawing heavily on Spencer and accepting his Lamarckian idea of the inheritance of acquired characteristics, Darwin sees particular emotional expressions as remnants of specific goal-directed habitual actions originally purposefully performed. Thus, in certain animals the expression of hostility will involve postural changes related to those which would prepare them for attack. In contrast, the third principle, that of "direct action," explained expressions that were not in any way derived from wilful or habitual action, but that were to be explained simply as the result of an overflowing of nerve force from a strongly excited sensorium.[36] But in contrast to these two principles, the second, the principle of antithesis, invoked the idea of a relation *between* expressions. It formed, as it were, a theory of a rudimentary morphology of expressive signs, and as such, invoked an idea much like that of Schelling's polarity.

As seen, some forms of expression such as those of hostility in carnivores such as dogs and cats are to be understood as related to intentional and purposeful actions. But the postures of dogs and cats expressing "an affectionate frame of mind" do not have an equivalent independent explanation; rather they are to be explained by the fact that they are related "antithetically" to the posture of their opposite—hostility. "This contrast in the attitudes and movements of these two carnivorous animals, under the same pleased and affectionate frame of mind, can be explained, as it appears to me, solely by their movements standing in complete antithesis to those which are naturally assumed, when these animals feel savage and are prepared either to fight or to seize their prey."[37] For Darwin these antithetical forms of expression had been selected because of the communicative advantage such clear contrasts afford: "The power of intercommunication between the members of the same community,—and with other species, between the opposite sexes, as well as between the young and the old,—is of the highest importance to them."[38]

It is to be noted that the opposition here does not simply apply to the external expressions involved but also applies *phenomenologically* to the quality of the experienced feelings as well: "When a *directly opposite state of mind* is induced, there is a strong and involuntary tendency to the performance of movements of a directly opposite nature."[39] Like many other nineteenth-century psychologists, Darwin did not share the twentieth-century tendency to leave aside the subjective side of mental life. Rather, sharing in the psychophysical parallelism of his generation, he just assumed that oppositions would be just as obvious subjectively as they were objectively: We directly experience the feelings of, say, anger and affection *as* opposites, just as we do happiness or sadness.

Qua phenomenological states Darwin's antitheses recall Goethe's appeal to color contrasts in his color theory, a classic application of the notion of polarity admired by both Schelling and Hegel.[40] The polarities of color phenomenology are now understood naturalistically in terms of the physiology of human vision. For example, it is now known that in humans different retinal cells have different "wavelength sensitivities" depending on which of the three retinal pigments they contain, and it has been shown how ingenious "wiring" of post-retinal neural circuits results in the "opponent processing" of color. Essentially, the outgoing signal from these circuits reflects more closely the chromatic *differences* within the incident light rather than its actual component frequencies. Thus, in the hierarchical processing of color vision, the signal leaving the post-retinal ganglia is a "second-order signal" that involves what Hughlings Jackson would have described as a "re-representation" of the original pattern of retinal stimulation.[41] And it seems that opponent processing might be a crucial factor in the expression of emotion in animals and humans as well. For example, the vocalizations involved in the sobbing and laughing of humans contain inverse breathing patterns, and the alarm calls of vervet monkeys also show pairs of opposed tokens.[42]

Opponent processing is a well-known phenomenon in psychology and is linked to the classical model of the conditioned reflex.[43] It would therefore appear to be relevant to the operations of indexical reference. It is present too, in the processing of human speech, but here there is a crucial difference. In human speech the smallest functional units relevant for the identification of words, "phonemes," show this type of opponent structure. Thus, for example, it is the opposition between the final consonants of "pig" and "pick" (/g/ and /k/) that allow speakers of English to understand these as tokens of different words. However, the compositionality of language (here manifested at the *word* level and commonly referred to by linguists as the principle of "double articulation") means that such individual phonemes are themselves *meaningless*. But all animal call systems lack such compositionality (at either word or sentence level) and for them it is the opposed sound units that are *themselves* meaningful. In fact, we might see this opponent structure as constitutive at the level of "iconic" reference that, as Deacon points out, is presupposed by the indexical reference of which such animals are capable.[44]

From an evolutionary perspective, we humans have not, of course, simply moved on from such call systems inherited from our prehuman ancestors to something better, language. Speech has not replaced smiling, frowning, laughing, and crying. Moreover, the intonational patterns of speech express affective states in similar ways and are controlled by what are essentially the same subcortical centers. But our added linguistic capacity means that we also now have the capacity to "re-represent" those affective states that would otherwise be restricted to being communicated nonlinguistically. This of course makes a world of difference. Such re-representation means that our felt states can be given new functional roles by becoming "poised for use" as premises in reasoning, action, and speech; and it is upon this fact that rests the degree of truth found in propositional attitude theories of emotion. But thinking that this is *all* there is to emotion misses the fact that the *"Aufhebung"* of affect in thought does not thereby simply deprive it of its more primitive functions.

Freud saw the secondary process not as replacing the primary process but as somehow developing out of it, and as a modification of it. Beneath the surface of rational thought, the dreamlike primary process still operated with its mad logic, which ignored the distinction between a representation and what it represented and, thereby, between appearance and reality. For Tomkins also, the affect system must still operate with its lack of discrimination of appearance and reality beneath the surface of conceptual reason. I must be able to be "rewarded" by the mere *anticipation* of some possible future state of affairs so that I can be motivated *now* to reason about how to achieve the *real* rewards of that as yet nonexistent state. This is why the appraisal theorists "who, like Everyman, cannot imagine feeling without an adequate 'reason'," have misunderstood the nature of affect. The having of affects cannot be grounded in reason because reasoning relies on the strictly *irrational* rewards and punishments provided by the affect system itself.

Does this mean that our reasoning is *determined* by something irrational? In linking the primary process with a *drive* such as the sex drive, Freud seemed to suggest just this. But, according to Tomkins, this concept rests on an unjustified conflation of drive with affect. A drive must motivate a *determinate* action (as hunger does eating), but the "objects" of the primary process are not "determinate," and their "achievements" do not depend on specific actions. By its very irrationality, by the fact that the primary process rewards by appearances rather than realities, the affect system cannot "steer" cognition, as it were, from without. Fichte, the early Schelling, and especially Hegel had all, in their own ways, argued that any Kantian "self-conceiving" apperceptive self must remain tethered to some immediate self-feeling, but that in itself such feeling was indeterminate and nonrepresentational. To play a determinate role *within* mental life, such states had to be brought into the space of theoretical and practical reason, incorporated into "representations" of the world—in Fichte's terminology, "posited." But there must have already been "something" subjective, intensive, and marked with positivity or negativity that could have been so assigned to such a role, a mental state not yet determinate, but rather merely "determinable."

It was such yet-to-be-determined feelings that, according to the idealists, lie at the origin of one's interpretative and evaluative construals of the world, one's "positings" in which, with the help of the public resources of language, an intelligible epistemic and practical relation to the world could be forged. But such a system must in some way be supported by a more basic layer of our embodied and enminded being in the world, a layer in which separation between an element and the functional role it plays has yet to emerge. Reason must navigate on a sea of biological and other natural forces that do not belong to it, but without which it could go nowhere. Affect is our most immediate awareness of the fact that we sail on such a sea.

Notes

Introduction

1. *"Le coeur a ses raisons, que la raison ne connaît point."* Blaise Pascal, *Pensées*, series II, §423.

2. This is especially true of Freud, whose ideas so clearly separated the twentieth century from the nineteenth.

3. J. G. Fichte, *The Science of Knowledge*, edited and translated by Peter Heath and John Lachs (Cambridge: Cambridge University Press, 1982), 247.

Chapter 1

1. William James, "What Is an Emotion?" in *Essays in Psychology* (Cambridge: Harvard University Press, 1983), 173–74.

2. From Walter B. Cannon, *Bodily Changes in Pain, Hunger, Fear and Rage*, 2d ed. (New York: Appleton, 1929), 351, quoted in Stanley Schachter and Jerome E. Singer, "Cognitive, Social, and Physiological Determinants of Emotional State," *Psychological Review* 69 (1962): 379. In recent physiological approaches to emotion there has been a significant swing back toward the Jamesian idea of distinct psychophysical differentiation of the emotions. See Richard J. Davidson, "Complexities in the Search for Emotion-Specific Physiology," in Paul Ekman and Richard J. Davidson, eds., *The Nature of Emotion: Fundamental Questions* (New York: Oxford University Press, 1994).

3. Gilbert Ryle, *The Concept of Mind* (London: Hutchinson, 1949).

4. J. J. C. Smart, "Sensations and Brain Processes," *Philosophical Review* 68 (1959): 141–56. Other important papers at about this time were U. T. Place, "Is Consciousness a Brain Process?" *British Journal of Psychology* 47 (1956): 44–50, and Herbert Feigl, "The 'Mental' and the 'Physical'," *Minnesota Studies in the Philosophy of Science* 2 (1958): 370–497.

5. D. M. Armstrong, *A Materialist Theory of the Mind* (New York: Humanities Press, 1968).

6. See, for example, Hilary Putnam, "Minds and Machines," in Alan Ross Anderson, ed., *Minds and Machines* (Englewood Cliffs, N.J.: Prentice-Hall, 1964), and Robert Cummins, "Functional Analysis," *Journal of Philosophy* 72 (1975): 741–65. Putnam, one of the first exponents of functionalism, was also one of its earliest critics. See, for example, *Representation and Reality* (Cambridge: MIT Press, 1989), chap. 5.

7. The most explicitly "representationalist" version of functionalism is perhaps that of Jerry Fodor. See, for example, his "Fodor's Guide to Mental Representation," in Stephen P. Stich and Ted A. Warfield, eds., *Mental Representation: A Reader* (Oxford: Blackwell, 1990).

8. See especially Howard Gardner, *The Mind's New Science: A History of the Cognitive Revolution* (New York: Basic Books, 1985).

9. Stanley Schachter and Jerome E. Singer, "Cognitive, Social, and Physiological Determinants of Emotional State," *Psychological Review* 69 (1962): 379–99.

10. The entry on "Emotion" in Richard L. Gregory's popular *Oxford Companion to the Mind* is fairly typical here—as in his characterization of the cognitive turn with the work of Stanley Schachter in the 1960s, which "cut the Gordian knot in which James's theory and the Jamesian critics had been entwined." Richard L. Gregory, ed., with the assistance of O. L. Zangwill, *The Oxford Companion to the Mind* (Oxford: Oxford University Press, 1987), 219–20. For a recent example of the strong cognitivist paradigm in psychology see Richard S. Lazarus and Bernice N. Lazarus, *Passion and Reason: Making Sense of Our Emotions* (New York: Oxford University Press, 1994).

11. See, for example, Anthony Kenny, *Action, Emotion, and Will* (London: Routledge and Kegan Paul, 1963); Robert Gordon, "The Aboutness of Emotions," *American Philosophical Quarterly* 2 (1974): 27–36; and Robert Solomon, *The Passions* (Notre Dame: University of Notre Dame Press, 1983).

12. Silvan Tomkins, "The Quest for Primary Motives: Biography and Autobiography of an Idea," in E. Virginia Demos, ed., *Exploring Affect: The Selected Writings of Silvan S. Tomkins* (Cambridge: Cambridge University Press, 1995).

13. See, for example, Davidson, "Complexities in the Search for Emotion-Specific Physiology."

14. Aaron Sloman, "Motives, Mechanisms, and Emotions," in Margaret A. Boden, ed., *The Philosophy of Artificial Intelligence* (Oxford: Oxford University Press, 1990), 232, emphasis added. See also the following: "Sometimes, in human beings, emotional states produce physiological disturbances too. . . . However, if X satisfied enough of the other conditions he could rightly be described as angry, even without any physical symptoms" (239).

15. Ibid., 242.

16. R. B. Zajonc, "On the Primacy of Affect," *American Psychologist* 39 (1984): 117–23. See also Zajonc's "Feeling and Thinking: Preferences Need No Inferences," *American Psychologist* 35 (1980): 151–75. The 1984 article is profitably read with the reply by appraisal theorist Richard Lazarus in the same volume—R. S. Lazarus, "On the Primacy of Cognition," *American Psychologist* 39 (1984): 124–29.

17. Ned Block, "On a Confusion about a Function of Consciousness," *Behavioral and Brain Sciences* 18 (1995): 227–87. Block's article is followed by a useful "peer review" of his distinction, with a final response by Block.

18. Thomas Nagel, "What Is It Like to Be a Bat?" *Philosophical Review* 83 (1974): 435–50, reprinted in his *Mortal Questions* (Cambridge: Cambridge University Press, 1979).

19. Block, "On a Confusion about a Function of Consciousness," 231.

20. Nicholas Humphrey, *A History of the Mind* (New York: Simon and Schuster, 1992). See also "Blocking Out the Distinction between Sensation and Perception: Superblindsight and the Case of Helen," in the peer review of Block, "On a Confusion about a Function of Consciousness."

21. In fact, for the most part, "appraisal theorists" were largely out of step with the majority position among cognitive scientists concerning the degree of access an individual had to his or her own mental states. Here Chomsky's ground-breaking notion of "linguistic competence" could be seen as setting the assumption that we *do not* have access to the cognitive processes underlying our various cognitive abilities.

22. Daniel C. Dennett, *The Intentional Stance* (Cambridge: MIT Press, 1987), x.

23. Michael Tye, *Ten Problems of Consciousness: A Representational Theory of the Phenomenal Mind* (Cambridge: MIT Press, 1995).

24. Fred Dretske, *Naturalizing the Mind* (Cambridge: MIT Press, 1995).

25. Louis C. Charland, "Feeling and Representing: Computational Theory and the Modularity of Affect," *Synthese* 105, 3 (1995): 273.

26. Claire Armon-Jones, *Varieties of Affect* (Toronto: University of Toronto Press, 1991). For another strong attack on the propositional attitude theory of emotions, this time from the point of view of a theory grounded in evolutionary biology, see Paul E. Griffiths, *What Emotions Really Are: The Problem of Psychological Categories* (Chicago: University of Chicago Press, 1997).

27. Armon-Jones, *Varieties of Affect*, 23. See also Cheshire Calhoun, "Cognitive Emotions?" in Cheshire Calhoun and Robert C. Solomon, *What Is an Emotion? Classic Readings in Philosophical Psychology* (New York: Oxford University Press, 1984), 331.

28. Armon-Jones, *Varieties of Affect*, 23–25. Among the "neo-cognitivist" critics of the standard cognitive model are Amélie Rorty, "Explaining Emotions," in A. Rorty, ed., *Explaining Emotions* (Berkeley: University of California Press, 1980); Michael Stocker, "Emotional Thoughts," *American Philosophical Quarterly* 24 (1987): 59–69; and Cheshire Calhoun, "Cognitive Emotions?" A similar idea used to counter the primacy of propositionally conceived cognition within the tradition of continental phenomenology is found in Heidegger's claims about the primordiality of the nonpropositional "hermeneutic as" to the conventionally propositional "apophantic as." Martin Heidegger, *Being and Time*, trans. J. Macquarrie and E. Robinson (Oxford: Blackwell, 1967).

29. Tye, *Ten Problems of Consciousness*, 104. For Charland it is this nonpropositionality of the relevant emotional representations that is behind the often commented-upon empirical fact that "rational persuasion is often basically helpless in altering and influencing affective reactions and attitudes." Charland, "Feeling and Representing," 285.

30. Tye, *Ten Problems of Consciousness*, 4. Block also thinks of propositional attitudes as paradigm "access conscious" states ("On a Confusion about a Function of Consciousness," 232). Tye, however, rejects Block's claim that phenomenally conscious states can be nonrepresentational.

31. Norton Nelkin, "Propositional Attitudes and Consciousness," *Philosophy and Phenomenological Research* 49 (1989): 423.

32. Ibid. Nelkin, too, while developing his argument in terms of an analysis of belief, seems to consider *all* propositional attitudes as nonphenomenal: "If I am right, then, in Nagel's sense, there is nothing that it is like to be a consciously believing, *desiring, hoping* or *fearing* being" (ibid., 430, emphasis added). But, we might respond, surely if Nagelian subjectivity applies to any mental state it applies to one such as *fear*—that is, if there is "something that it is like" to be anything, there is "something that it is like" to be frightened. But if this is right and Nelkin and Tye are right about propositional attitudes, then it must be the case that we cannot equate being frightened with being in such an attitude—with being "fearful that. . . ."

33. Antonio R. Damasio, *Descartes' Error: Emotion, Reason, and the Human Brain* (New York: G. P. Putnam's Sons, 1994).

34. Joseph LeDoux, *The Emotional Brain: The Mysterious Underpinnings of Emotional Life* (New York: Simon and Schuster, 1996).

35. Le Doux, *The Emotional Brain*, 40. Damasio, after quoting part of the passage from James quoted at the start of this chapter notes: "With these words, well ahead of both his time and ours, I believe William James seized upon the mechanism essential to the understanding of emotion and feeling" (*Descartes' Error*, 129).

36. Le Doux, *The Emotional Brain*, 40.

37. Ibid., 295. Concerning this last point, it would seem that the voluntary activation of those muscles involved in smiling actually produces positive affects. For a review of the experimental evidence, see P. K. Adelmann and R. Zajonc, "Facial Efference and the Experience of Emotion," *Annual Review of Psychology* 40 (1989): 348–80.

38. He compares situations where a response is "reflex" and without deliberation (action taken to avoid a falling rock, for example) with those that do involve deliberation (choosing a career, deciding whom to marry or vote for, and so on). Decision in the latter group of examples may involve conscious forms of inference and weighings of costs and benefits, but, stressing the continuity with reflex actions, Damasio argues for the relevance of "gut feelings" here as well. Envisaged bad outcomes, he suggests, "however fleetingly," bring about unpleasant feelings. These reactions narrow the scope of alternatives to be reasoned about and chosen between. Such a "biasing device" assists deliberations "by highlighting some options . . . and eliminating them rapidly from subsequent consideration" (Damasio, *Descartes' Error*, 174).

39. Daniel C. Dennett, *Kinds of Minds: Towards an Understanding of Consciousness* (London: Weidenfeld and Nicolson, 1996), 69.

40. Ibid., 71.

41. Ibid., 75. Dennett has in mind here facts such as the way neurotransmitters operate at the synaptic junctions between neurons.

42. More directly he sees this fact about the nervous system behind the "intuitively appealing claim often advanced by critics of functionalism: that it really does matter what you make a mind out of" (ibid., 76).

43. Ibid., 75.

44. Charland, "Feeling and Representation," 292–93.

45. Ibid., 276.

46. Tye, *Ten Problems of Consciousness*, 126.

47. Ibid., 101.

48. Ibid., 100.

49. Fred Dretske, "The Intentionality of Cognitive States," in Peter A. French, Theodore E. Uehling, Jr., and Howard K. Wettstein, eds., *Midwest Studies in Philosophy*, vol. V, reprinted in David M. Rosenthal, *The Nature of Mind* (New York: Oxford University Press, 1991). Page numbers are from the latter volume.

50. Ibid., 359.

51. But they could be taken as representational by, say, some other part of the system, just as a trees rings can be taken as representing age.

52. Smart, "Sensations and Brain Processes."

53. A version of psychophysical parallelism or "dual aspect theory" has recently been revived by David J. Chalmers, in *The Conscious Mind: In Search of a Fundamental Theory* (Oxford: Oxford University Press, 1996).

54. Block, "On a Confusion about a Function of Consciousness," 234.

55. The issue of the nonconscious processing of emotional states has raised the questions of what it is for a feeling to be conscious. LeDoux speculates that conscious emotional experiences "are probably created the same way that other conscious experiences are—by the establishment of a conscious representation of the workings of underlying processing systems" (*The Emotional Brain*, 269).

56. Nicholas Humphrey, *A History of the Mind* (London: Chatto and Windus, 1992), 26.

57. Humphrey, *A History of the Mind*, 58–61. Bach-y-Rita's work was reported in his book *Brain Mechanisms in Sensory Substitution* (London: Academic Press, 1972).

58. Paul Bach-y-Rita, *Brain Mechanisms in Sensory Substitution*, quoted in Humphrey, *A History of the Mind*, 60–61.

Chapter 2

1. Walter B. Cannon, *Bodily Changes in Pain, Hunger, Fear, and Rage*, 2d ed. (New York: Appleton, 1929), extracts republished in Cheshire Calhoun and Robert C. Solomon, eds., *What Is an Emotion? Classical Readings in Philosophical Psychology* (Oxford: Oxford University Press, 1984), 143.

2. "An object falls on a sense-organ and is apperceived by the appropriate cortical centre; or else the latter, excited in some other way, gives rise to an idea of the same object. Quick as a flash, the reflex currents pass down through their pre-ordained channels, alter the condition of muscle, skin and viscus; and these alterations, apperceived like the original object, in as many specific portions of the cortex, combine with it in consciousness and transform it from an object-simply-apprehended into an object-emotionally-felt." James, "What Is an Emotion?" 184.

3. See "*Whilst part of what we perceive comes through our senses from the object before us, another part* (and it may be the larger part) *always comes . . . out of our own head.*" William James, *The Principles of Psychology*, (New York: Dover, 1950), vol. II, 103.

4. See Robert J. Richards's account of Spencer's "evolutionary Kantianism" in *Darwin and the Emergence of Evolutionary Theories of Mind and Behavior* (Chicago: University of Chicago Press, 1987), 285–91. Richards discusses James's complex relation to Herbert Spencer at 422–30.

5. James, "What Is an Emotion?" 170.

6. Ibid., 170–71.

7. John Dewey, "The Reflex Arc Concept in Psychology," *Psychological Review* 3 (1896): 363.

8. C. W. Slack, "Feedback Theory and the Reflex Arc Concept," *Psychological Review* 62 (1955): 263–67. In "James, Dewey, and the Reflex Arc," *Journal of the History of Ideas* 32 (1971), 566, D. C. Phillips links Dewey's understanding to his earlier Hegelianism. Charles Darwin, *The Expression of the Emotions in Man and Animals* (Chicago: University of Chicago Press, 1965), first published 1872.

9. James, "What Is an Emotion?" 175.

10. Ibid.

11. Ibid.

12. As Irvin Rock points out, James makes no distinction between *perception* and *recognition*, "A Look Back at William James's Theory of Perception," in Michael G. Johnson and Tracy B. Henley, eds., *Reflections on the Principles of Psychology: William James after a Century* (Hillsdale, N.J.: Lawrence Erlbaum Associates, 1990), 202.

13. In 1894 James replied to actual critics voicing the same "cognitive" objection. See, for example, his reply to one Dr. W. L. Worchester in "The Physical Basis of Emotion," in *Essays in Psychology* (Cambridge, Mass.: Harvard University Press, 1983), 301.

14. Edwin Clarke and L. S. Jacyna, *Nineteenth-Century Origins of Neuroscientific Concepts* (Berkeley: University of California Press, 1987), 124.

15. Ibid., 129.

16. Ibid., 132.

17. On this fascinating group see, for example, Philip F. Rehbock, *Philosophical Naturalists: Themes in Early-Nineteenth-Century British Biology* (Madison: University of Wisconsin Press, 1983), chaps. 1–3; Dov Ospovat, *The Development of Darwin's Theory: Natural History, Natural Theology, and Natural Selection, 1838–1859* (Cambridge: Cambridge University

Press, 1981), chap. 1; and Edward S. Reed, *From Soul to Mind: The Emergence of Psychology from Erasmus Darwin to William James* (New Haven: Yale University Press, 1997), esp. chap. 4. For fuller accounts of the intersecting careers of Laycock and Carpenter, see Michael Barfoot, "Introduction" to Michael Barfoot, ed., *"To Ask the Suffrages of the Patrons": Thomas Laycock and the Edinburgh Chair of Medicine, 1855* (London: Welcome Institute for the History of Medicine, 1995), and J. Estlin Carpenter, "Introductory Memoir" to William B. Carpenter, *Nature and Man: Essays Scientific and Philosophical* (London: Kegan Paul, Trench & Co., 1888).

18. Clarke and Jacyna, *Nineteenth-Century Origins*, 141–47; Kenneth Dewhurst, *Hughlings Jackson on Psychiatry* (Oxford: Sandford, 1982); and Anne Harrington, *Medicine, Mind, and the Double Brain: A Study in Nineteenth-Century Thought* (Princeton: Princeton University Press, 1987), chap. 7. Jackson's essays and lectures have been collected in *Selected Writings of John Hughlings Jackson*, 2 vols. (London: Hodder and Soughton, 1931).

19. Theodor Meynert, *Psychiatrie: Klinik der Erkrankungen des Vorderhirns begründet auf dessen Bau, Leistungen und Ernährung* (Vienna: Wilhelm Braumüller, 1884). An English translation by B. Sachs was published by G. P. Putnam's Sons (New York and London) in 1884 as *Psychiatry: A Clinical Treatise on Diseases of the Fore-Brain Based upon a Study of Its Structure, Function, and Nutrition.*

20. James, *Principles of Psychology*, I. 27.

21. Ibid., I. 72.

22. Ibid., I. 80.

23. Ibid., II, chaps. 24, 25, and 26. In the context of his discussion of the will, James uses Carpenter's notion of reflex "ideo-motor action" (ibid., 522).

24. James does not mention Jackson in his brief discussion of aphasia in "The Functions of the Brain," *Principles of Psychology*, I. Chap. 2.

25. Meynert, *Psychiatry*, 255–56. On this and other aspects of Meynert's relation to Darwinian theory, see Lucille B. Ritvo, *Darwin's Influence on Freud: A Tale of Two Sciences* (New Haven: Yale University Press, 1990), 172–78.

26. Jackson, *Selected Writings*, 2, 53.

27. From J. H. Jackson, "On Some Implications of Dissolution of the Nervous System," *Medical Press and Circular* 2 (1882): 411; quoted in Harrington, *Medicine, Mind, and the Double Brain*, 212.

28. James, *Principles of Psychology*, I. 24, footnote.

29. Ibid., II. 453.

30. Dewey, "The Reflex Arc Concept in Psychology," 365.

31. In this century philosophers have tended to use the former term, while in the nineteenth century the latter predominates. Given that the nineteenth century represents the high point of the "parallelist" doctrine, I will use its terminology.

32. From J. C. Reil, *Rhapsodien*, quoted in Theodore Ziolkowski, *German Romanticism and Its Institutions* (New Haven: Yale University Press, 1990), 183. Clarke and Jacyna note Reil's support of the "down-up" organization of the central nervous system, which effectively replaced the "top-down" model that had predominated since Galen. Reil "recognized that, to elucidate the organization of this organ, 'one can go forwards from the spinal cord to the brain or backwards from the brain to the spinal cord'." *Nineteenth-Century Origins of Neuroscientific Concepts*, 52. See also, LeeAnn Hansen, "From Enlightenment to *Naturphilosophie*: Marcus Herz, Johann Christian Reil, and the Problem of Border Crossings," *Journal of the History of Biology* 26 [1993]: 39–64.

33. Gustav T. Fechner, *Elemente der Psychophysik* (Leipzig: Brietkopf und Härtel, 1860). Translated as *Elements of Psychophysics*, vol. 1, translated by H. E. Adler (New York: Holt, Rinehart and Winston, 1966).

34. Later, a similar view was developed by Ernst Mach. See for example, Ernst Mach, "Lectures on Psychophysics—Conclusion" (1863), translated in J. Blackmore, ed., *Ernst Mach—A Deeper Look* (Dordrecht: Kluwer, 1992).

35. Fechner, *Elements of Psychophysics*, 2–3.

36. Rosemary Ashton, *The German Idea: Four English Writers and the Reception of German Thought* (Cambridge: Cambridge University Press, 1980), 130. See also her *G. H. Lewes: A Life* (Oxford: Oxford University Press, 1991). On Lewes's influence on nineteenth-century thought, see Reed, *From the Soul to the Mind*, chap. 8; and on the fascinating account of Lewes's introduction to the ideas of Spinoza in a London pub in the 1830s, see ibid., 261. Lewes was, of course, first the companion and then husband of Mary Ann Evans/George Eliot. Lewes's psychological writings include *The Physical Basis of Mind* (London: Trübner, 1877) and *Physiology of Common Life* (London: Blackwood, 1859–1860).

37. See, for example, Herbert Spencer, *Principles of Psychology*, 2d ed. (London: Williams and Norgate, 1870), 99–100, where he discusses the "correlation" of feelings and nervous excitation.

38. James, *Principles of Psychology*, I. 29–30.

39. Dewey, "The Reflex Arc Concept in Psychology," 365.

40. James, *Principles of Psychology*, I. 30.

41. It has been pointed out that it was the application of the parallelist picture in Fechner's psychophysics that James objected to (Gerald E. Myers, *William James, His Life and Thought* [New Haven: Yale University Press, 1986], 107–8). In contrast, James seemed quite happy with the extremely odd metaphysical picture, involving world souls and the like, that Fechner held to as a metaphysico-religious belief that was compatible with the science that he practiced.

42. William James, "The Function of Cognition," reprinted in *Pragmatism and The Meaning of Truth* (Cambridge, Mass.: Harvard University Press, 1975), 179–98.

43. This becomes very explicit in the later paper "Does 'Consciousness' Exist?" in *Essays in Radical Empiricism and A Pluralistic Universe* (New York: Longmans, 1947): "For twenty years past I have mistrusted 'consciousness' as an entity; for seven or eight years past I have suggested its non-existence to my students, and tried to give them its pragmatic equivalent in realities of experience. It seems to me that the hour is ripe for it to be openly and universally discarded" (3).

44. Myers, *William James*, 57–58.

45. In recent work this "direct realist" account of perception has been prominently represented by the work of J. J. Gibson and those influenced by him. On the work of Gibson see Edward S. Reed, *James J. Gibson and the Psychology of Perception* (New Haven: Yale University Press, 1988).

46. William James, "The Tigers in India," first published in the *Psychological Review*, vol. II (1895), and reprinted in *Pragmatism and The Meaning of Truth*, 199–202.

47. Ibid, 202. As we can see from James's discussion, it is not as if the direct presentationalist approach has *no place* for the notion of mental representation. Rather, what it denies is that the notions of intentionality or consciousness can be entirely *explained* in terms of the notion of representation.

48. Ibid., 202, note.

49. James, "Does 'Consciousness' Exist?" 11, 12, 13.

50. William James, "The Place of Affectional Facts in a World of Pure Experience," in *Essays in Radical Empiricism*, 142.

51. Ibid., 143.

52. Ibid., 153.

53. Ibid.

54. James was of course, firmly opposed to Kant's "transcendentalism" by which he meant Kant's reference to a conceivable (but nonknowable) world of things in themselves beyond that of pure experience. Ghostly pseudo-worlds such as these are precisely what develop when one hypostatizes "concepts" beyond their actual status as pragmatically motivated secondary substitutions for the particular contents of experience.

55. Myers, *William James*, 49.

Chapter 3

1. Ned Block, "On a Confusion about a Function of Consciousness," *Behavioral and Brain Sciences* 18 (1995): 231.

2. See in particular Sigmund Freud, *The Interpretation of Dreams*, chap. 7; "Formulations on the Two Principles of Mental Functioning"; "A Note on the Unconscious in Psychoanalysis"; "The Unconscious"; "The Ego and the Id"; and *New Introductory Lectures on Psychoanalysis*, lecture 29. All page numbers given here are to *The Standard Edition of the Complete Psychological Works of Sigmund Freud*, translated and edited by James Strachey (London: Hogarth, 1953–1973), and indicated by *SE* followed by volume and page numbers.

3. Frank J. Sulloway, *Freud, Biologist of the Mind: Beyond the Psychoanalytic Legend* (New York: Basic Books, 1979).

4. Sigmund Freud, *The Interpretation of Dreams*, SE, 5. 538.

5. Freud published a monograph on aphasia, *Zur Auffassung der Aphasien*, in 1891, not translated into English until 1953 as *On Aphasia* (New York: International University Press, 1953). Both the significance for Freud's later work of his aphasia monograph and the influence of Jackson's thought on Freud have recently received considerable attention. See, for example, A.-M. Rizzuto, "The Origins of Freud's Concept of Object Representation (*Objektvorstellung*) in His Monograph 'On Aphasia,' Its Theoretical and Technical Importance," *International Journal of Psychoanalysis* 71 (1990): 241–48, and "Freud's Speech Apparatus and Spontaneous Speech," *International Journal of Psychoanalysis*, 74 (1993): 113–27; John Forrester, *Language and the Origins of Psychoanalysis* (London: Macmillan, 1980); and Anne Harrington, *Medicine, Mind, and the Double Brain: A Study in Nineteenth-Century Thought* (Princeton: Princeton University Press, 1987).

6. Freud, *On Aphasia*, 50 and 52.

7. Ibid., 53–54.

8. Freud, *The Interpretation of Dreams, SE*, 5. 538.

9. Freud, *On Aphasia*, 57.

10. Ibid., 55.

11. Sigmund Freud, "The Unconscious," *SE*, 14. 168.

12. This had been most explicitly developed in the early and abandoned "Project for a Scientific Psychology" (1895), but is also explicitly present in *The Interpretation of Dreams*, chap. 7: "Hypotheses, whose justification must be looked for in other directions, tell us that at first the apparatus's efforts were directed towards keeping itself so far as possible free from stimuli; consequently its first structure followed the plan of a reflex apparatus, so that any sensory excitation impinging on it could be promptly discharged along a motor path" (*SE*, 5. 565). Freud returned to this theme in his formulations in 1920 of the "death drive" in "Beyond the Pleasure Principle." G. T. Fechner had developed the principle of constancy in *Einige Ideen zur Schöpfungs- und Entwicklungsgeschichte der Organismen* (Leipzig: Breitkopf und Härtel, 1873).

13. This idea can also be traced back to earlier German psychiatric thought as Wilhelm Griesinger had conceived of the contents of the experience of dreams and psychoses as wish fulfilments.

14. James, *The Principles of Psychology*, 2 vols. (New York: Dover, 1950), 1. 24–7.

15. Freud, *The Interpretation of Dreams, SE* 5. 565–66.

16. Ibid., 566.

17. Moreover, they used their propensity to produce hallucinatory forms of gratification as the basis for their own type of imagistic thinking in which one's actions in the world could be "rehearsed."

18. Freud, *On Aphasia,* 49–50.

19. LeDoux (Joseph LeDoux, *The Emotional Brain: The Mysterious Underpinnings of Emotional Life* [New York: Simon and Schuster, 1996]), mentions the work of J. W. Papez in the 1930s ("A Proposed Mechanism of Emotion," *Archives of Neurology and Psychiatry* 79 [1937]: 217–24) and of Paul MacLean in the 1940s ("Psychosomatic Disease and the 'Visceral Brain.' Recent Developments Bearing on the Papez Theory of Emotion," *Psychosomatic Medicine* 11 [1949]: 338–53) as putting forward such a view. But as Horace Magoun has shown (in "Evolutionary Concepts of Brain Function Following Darwin and Spencer," in Sol Tax, (ed.,) *Evolution after Darwin, Vol. II, The Evolution of Man* [Chicago: Chicago University Press, 1960], 187–209), it was put forward by a variety of theorists toward the end of the nineteenth century, including Jackson, Freud, and Pavlov.

20. Again there is contemporary support for this view. On the increased sensitivity in monkeys and apes to emotional perceptual cues that come from the left side of their visual field (and are hence registered in the right visual cortex), see Robin Dunbar, *Grooming, Gossip, and the Evolution of Language* (London: Faber and Faber, 1996), 138. On the lateralization of emotional responses in humans see, for example, Richard J. Davidson, "Cerebral Asymmetry, Emotion, and Affective Style," in Richard J. Davidson and Kenneth Hugdahl, eds., *Brain Asymmetry* (Cambridge, Mass.: MIT Press, 1995).

21. Harrington, *Medicine, Mind, and the Double Brain,* 216. Herder's essay is translated in *On the Origin of Language,* translated by John H. Moran and Alexander Gode with an introduction by Alexander Gode (Chicago: University of Chicago Press, 1966).

22. See, for example, R. G. Goldstein, "The Higher and Lower in Mental Life: An Essay on J. Hughlings Jackson and Freud," *Journal of the American Psychoanalytic Association,* 43 (1995): 495–515, as well as Rizzuto, "The Origins of Freud's Concept of Object Representation," and "Freud's Speech Apparatus and Spontaneous Speech."

23. C. G. Jung, "Die Psychopathologische Bedeutung des Assoziationsexperimentes," *Archiv für Kriminal-Anthropologie und Kriminalistik* 22 (1906): 145–62. Francis Galton's "Antechamber of Consciousness" and "Psychometric Experiments" were reprinted in his *Inquiries into Human Faculty* (London: Dent, 1907). On the history of this development see Henri F. Ellenberger, *The Discovery of the Unconscious: The History and Evolution of Dynamic Psychiatry* (New York: Basic Books, 1970), 691–94.

24. Galton, "Psychometric Experiments," 134.

25. Ibid.

26. Ibid., 135 and 138.

27. See Ellenberger, *Discovery of the Unconscious,* 202–15. On the importance of hypnotism for the "unofficial" psychology of the nineteenth century, see Edward S. Reed, *From Soul to Mind: The Emergence of Psychology from Erasmus Darwin to William James* (New Haven: Yale University Press, 1997), Chap. 6.

28. Gotthilf Heinrich Schubert), *Die Symbolik des Traumes,* Faksimiledruck nach der Ausgabe von 1814, mit einem Nachwort von Gerhard Sauder (Heidelberg: Lambert Schneider, 1968); Freud, *The Interpretation of Dreams, SE,* 4.63 and 5.352.

29. Schubert, *Die Symbolik des Traumes,* 1–2.

30. In the 1840s a group of young medical scientists, Emil du Bois Reymond, Hermann Helmholtz, Ernst Brücke, and Carl Ludwig attempted to banish the romantic movement within the medical sciences and advocated a reductionist approach to physiology, the so-called "bio-physics program." Significantly their antiromanticism and distrust of notions suggestive of a "spinal soul" led toward the reinstatement of a more tra-

ditional Cartesian approach to the mind against the earlier picture based on the cerebral reflex.

31. Lancelot Law Whyte, *The Unconscious before Freud* (London: Associated Book Publishers, 1967), 163. Eduard von Hartmann, *Philosophie des Unbewussten: Versuch einer Weltanschauung* (Berlin: Carl Duncker, 1869). Translated by W. C. Coupland as *The Philosophy of the Unconscious* (London: English and Foreign Photosophical Library, 1884).

32. E. J. Buzzard, quoted in Michael Swash, "J. Hughlings Jackson: A Sesquicentennial Tribute," *Journal of Neurology, Neurosurgery and Psychiatry* 49 (1986): 983.

33. On this movement see especially Philip F. Rehbock, *Philosophical Naturalists: Themes in Early-Nineteenth-Century British Biology*, (Madison: University of Wisconsin Press, 1983).

34. Adrian Desmond and James Moore, *Darwin* (London: Michael Joseph, 1991), 394; Magoun, "Evolutionary Concepts of Brain Function"; and H. J. Critchley, "Hughlings Jackson: The Man and His Time," *Archives of Neurology* 43, 5 (1986): 435. Von Baer has been described by Timothy Lenoir as belonging to the "teleomechanist" tradition of early-nineteenth-century German biology, a tradition that conceived of biological explanation on the model that Kant had sketched in his *Critique of Judgment*, rather than that of "nature philosophy." Timothy Lenoir, *The Strategy of Life: Teleology and Mechanics in Nineteenth-Century German Biology* (Dordrecht: Reidel, 1982), 72–95; see also Robert J. Richards's criticism of this in *The Meaning of Evolution: The Morphological Construction and Ideological Reconstruction of Darwin's Theory* (Chicago: University of Chicago Press, 1992), 59, note 81.

35. In fact, much attention has been given to the possible influence of the earlier idealist tradition on Darwin himself. See, for example, Phillip Sloan's introductory essay to Phillip R. Sloan, ed., *Richard Owen: The Hunterian Lectures in Comparative Anatomy, May–June, 1937. With an Introductory Essay and Commentary* (Chicago: University of Chicago Press, 1992).

36. We have already seen how James could entertain the possibility that subcortical levels of neuronal connection might have their own dissociated "consciousness" to which "cortical consciousness" was blind.

37. Freud, "The Unconscious," *SE*, 14. 171.

38. Among contemporary philosophers, David Armstrong supports a view somewhat like the one expressed by Freud here. See David M. Armstrong, *The Nature of Mind and Other Essays* (Ithaca: Cornell University Press, 1980). In contrast, Kant is usually understood as invoking the necessity of a "higher-order" *thought* rather than some quasi-perceptual experience. See, Rocco J. Gennero, *Consciousness and Self-Consciousness: A Defense of the Higher-Order Thought Theory of Consciousness* (Amsterdam: John Benjamins, 1995).

39. Kant, *Critique of Pure Reason*, translated by N. Kemp Smith (New York: Macmillan, 1929), A78 B103, emphasis added.

40. See Chapters 5 and 6.

41. Immanuel Kant, *Metaphysical Foundations of Natural Science*, translated by James W. Ellington, in Immanuel Kant, *Philosophy of Material Nature* (Indianapolis: Hackett, 1985), 8.

42. Furthermore, as Kant argued in the "Refutation of Idealism" added to the second edition of the *Critique of Pure Reason*, such objects of internal sense, although they themselves do not exist in space, nevertheless depend for their content on the presence to the mind of the enduring spatial objects of outer sense.

43. Kant, *Critique of Pure Reason*, B208.

44. F. W. J. von Schelling, *System of Transcendental Idealism*, translated by Peter Heath with an Introduction by Michael Vater (Charlottesville: University Press of Virginia, 1978), 100.

45. See, for example, Norton Nelkin, "Unconscious Sensations," *Philosophical Psychology* 2 (1989): 129–41, and Thomas Natsoulas, "Conscious Perception and the Paradox of 'Blind-Sight,' " in G. Underwood, ed., *Aspects of Consciousness, vol. 3, Awareness and Self-Consciousness* (London: Academic Press, 1982).

46. Immanuel Kant, *Anthropology from a Pragmatic Point of View*, translated by Victor Lyle Dowdell with an Introduction by Frederick P. Van De Pitte (Carbondale: Southern Illinois University Press, 1978).

47. Ibid.: "Many people are unhappy because they cannot engage in abstraction. Many a suitor could make a good marriage if he could only shut his eyes to a wart on his sweetheart's face or to a gap where teeth are missing" (14–15).

48. Ibid., 14.

49. Ibid., 25 and 26.

50. Schelling, *System of Transcendental Idealism*, 61.

51. Freud, "Formulations on the Two Principles of Mental Functioning," *SE*, 12. 220.

52. Ibid., 219.

53. Ibid., 220.

54. Ibid., 220–21.

55. On Herbart's development of Kantian psychology see Katherine Arens, *Structures of Knowing: Psychologies of the Nineteenth Century* (Dordrecht: Kluwer, 1989), and Gary C. Hatfield, *The Natural and the Normative: Theories of Spatial Perception from Kant to Helmholtz* (Cambridge, Mass.: MIT Press, 1990).

56. According to John Kerr, Bleuler and Jung interpreted Freud in this Kantian way. John Kerr, *A Most Dangerous Method: The Story of Jung, Freud, and Sabina Spielrein* (New York: Knopf, 1993), 110.

57. Sigmund Freud, "The Unconscious," *SE*, 14. 177. From the psychophysical point of view, the affect involved is just the subjective aspect of the physiological reality which is the instinct. This then is why Freud states that "the antithesis of conscious and unconscious is not applicable to instincts" (ibid).

58. See especially the discussion by Jonathan Lear in *Love and Its Place in Nature: A Philosophical Interpretation of Freudian Psychoanalysis* (New York: Farrar, Straus and Giroux, 1990), chap. 2. Adopting a "propositional attitude" approach to emotions Lear's Freudianism allows him to tie this articulation of emotion to the idea of underlying "proto-emotions" existing at the level of the unconscious and subject to its logic rather than that of language.

59. Freud, "The Unconscious," *SE*, 14. 178.

60. Ibid., 202.

61. Freud, "The Ego and the Id," *SE*, 19. 20.

62. Kant, *Anthropology from a Pragmatic Point of View*, 85.

63. Ibid., 12.

64. In J. F. Herbart, *Psychologie als Wissenschaft neu gegründet auf Erfahrung, Metaphysik und Mathematik*, in Karl Kehrbach, ed., *Sämtliche Werke* (Langensalza: Herman Beyer & Söhne, 1887-1912), vol. 5.

65. Kant, *Critique of Pure Reason*, A166 B208.

66. Freud, "Beyond the Pleasure Principle," *SE*, 18. 27.

67. Ibid., 27–28.

68. Ibid., 28.

69. "A Note Upon the 'Mystic Writing-Pad'," *SE*, 19. 231. Again Freud links such perceptual "sampling" to the concept of time: "I further had a suspicion that this discontinuous method of functioning of the system *Pcpt.-Cs.* lies at the bottom of the origin of the concept of time" (ibid).

70. Freud, "Beyond the Pleasure Principle," *SE*, 18. 28–29.

71. Ibid., 29.

72. Klaus Doerner, *Madmen and the Bourgeoisie: A Social History of Insanity and Psychiatry* (Oxford: Blackwell, 1981), 234–35. For a general account of the role played by Schellingian nature philosophy in the changes in German medicine at the turn of the nineteenth century

see Thomas H. Broman, *The Transformation of German Academic Medicine, 1750–1820* (Cambridge: Cambridge University Press, 1996), 90–101.

73. See S. R. Morgan, "Schelling and the Origin of His Nature Philosophy," in Andrew Cunningham and Nicholas Jardine, eds., *Romanticism and the Sciences* (Cambridge: Cambridge University Press, 1990).

74. See, for example, Albert Béguin, *L'Ame romantique et le rêve* (Paris: Corti, 1939).

75. Nicholas Jardine, "*Naturphilosophie* and the Kingdoms of Nature," in N. Jardine, J. A. Secord, and E. C. Spary, eds., *Cultures of Natural History* (Cambridge: Cambridge University Press, 1996), 230.

76. Uwe Henrik Peters, *Studies in German Romantic Psychiatry: Justinus Kerner as a Psychiatric Practitioner, E. T. A. Hoffmann as a Psychiatric Theorist* (London: Institute of German Studies, 1990).

77. Heinrich Straumann, *Justinus Kerner und der Okkultismus in der Deutschen Romantik* (Leipzig: Verlage der Münster-Presse, Horgen-Zürich, 1928). Straumann lists as the three aspects of Schellingian irrationalism that structure the Seeress episode the ideas of polarity, the universal nexus in nature, and the "felt" world (ibid., 28).

78. See Ellenberger, *Discovery of the Unconscious*, for a detailed account of the relation of the ideas of Jung to earlier romantic philosophy and medicine. On the vexed question of the Freud–Jung relationship, see John Kerr, *A Most Dangerous Method*; the accounts of Freud's and Jung's respective papers are at 272–76. See also Forrester, *Language and the Origins of Psychoanalysis*, 101. A very critical account of Jung is offered by Richard Noll in *The Jung Cult: Origins of a Charismatic Movement* (Princeton: Princeton University Press, 1994), which, however, oversimplifies the complexities of the nature philosophy movement and its relation to science. For a more balanced account of the relations of nineteenth-century "spiritualism" to the development of psychology, see Reed, *From Soul to Mind*.

79. Here Jung could feel himself to have some sort of scientific support in the ideas of Ernst Haeckle, a well-known biologist and popularizer of Darwinian theory, who advocated the idea that "ontogeny recapitulated phylogeny" as well as the inheritance of acquired characteristics.

80. Freud, *New Introductory Lectures on Psychoanalysis, SE*, 22. 80. Here I have followed Jonathan Lear in rendering *Ich* straightforwardly as "I" rather than "Ego," as in the Strachey translation. As Lear points out, Freud here seems concerned not so much with "the replacement of psychic agencies, but with a transformation of the relation in which I stand to my own instinctual life" (*Love and Its Place in Nature*, 168).

Chapter 4

1. See especially Patricia Kitcher, *Kant's Transcendental Psychology* (Oxford: Oxford University Press, 1990), and Andrew Brook, *Kant and the Mind* (Cambridge: Cambridge University Press, 1994).

2. Ralf Meerbote, "Kant's Functionalism," in J. C. Smith, ed., *Historical Foundations of Cognitive Science* (Dordrecht: Kluwer, 1990). Daniel Dennett had remarked on the kinship between cognitive science and the philosophy of Kant in "Artificial Intelligence as Philosophy and Psychology," in *Brainstorms* (Montgomery, Vt.: Bradford Books, 1978). Onora O'Neill (in "Transcendental Synthesis and Developmental Psychology," *Kant Studien* 75 [1984]: 149–67) has also interpreted Kant's transcendental psychology as a type of developmental cognitive psychology.

3. Patricia Kitcher, *Kant's Transcendental Psychology* (Oxford: Oxford University Press, 1990), 77–79.

4. Ralf Meerbote, "Apperception and Objectivity," *The Southern Journal of Philosophy* 25, suppl. (1987): 118.

5. Brook, *Kant and the Mind*, 12.

6. See, for example, J. A. Fodor, *The Modularity of Mind* (Cambridge, Mass.: MIT Press, 1983).

7. Wilfrid Sellars, "Metaphysics and the Concept of a Person," in Karel Lambert, ed., *The Logical Way of Doing Things* (New Haven: Yale University Press, 1969); "This I or He or It (This Thing) Which Thinks," Presidential Address to the Eastern Division of the APA, 1970; and "Kant's Views on Sensibility and Understanding," *The Monist* 51 (1967): 463–91.

8. See, for example, Hud Hudson, *Kant's Compatibilism* (Ithaca: Cornell University Press, 1994).

9. See Brook, *Kant and the Mind*, 13.

10. For a broadly Kantian characterization of such an account of consciousness, see Rocco J. Gennaro, *Consciousness and Self-Consciousness: A Defense of the Higher-Order Thought Theory of Consiousness* (Amsterdam: John Benjammins, 1995).

11. Indeed, such a "two-aspect" reading of Kant gives his account of experience a distinctly Jamesian, "direct realist," look.

12. See, for example, David E. Leary, "The Philosophical Development of the Concept of Psychology in Germany, 1780–1850," *Journal of the History of the Behavioural Sciences* 14 (1978): 113–21; Katherine Arens, *Structures of Knowing: Psychologies of the Nineteenth Century,* (Dordrecht: Kluwer, 1989); and Gary C. Hatfield, *The Natural and the Normative: Theories of Spatial Perceptions from Kant to Helmholtz* (Cambridge, Mass.: MIT Press, 1990).

13. Peter Strawson, *The Bounds of Sense: An Essay on Kant's "Critique of Pure Reason"* (London: Methuen, 1966).

14. Immanuel Kant, *Critique of Pure Reason,* translated by Norman Kemp Smith (New York: Macmillan, 1929), B 131.

15. Strawson, *The Bounds of Sense*, 28–29.

16. Kitcher, *Kant's Transcendental Psychology*, 13, 9–11.

17. Kitcher departs from convention here by translating Kant's *Vorstellung* not as "representation" but as "cognitive state," keeping "representation" for Kant's *Erkenntnis* (which would more commonly be translated as "cognition"). Kitcher's departure from tradition is motivated by the idea that not all Kantian *Vorstellungen* "represent," but can, rather, play contributory roles *within* representations (*Kant's Transcendental Psychology,* 244, note 12).

18. Strawson, *The Bounds of Sense*, 98, quoted in Kitcher, *Kant's Transcendental Psychology*, 92.

19. Kitcher, *Kant's Transcendental Psychology*, 128.

20. See, for example, Henry E. Allison, "Kant's Refutation of Materialism," and "Apperception and Analyticity in the B-Deduction," both in his *Idealism and Freedom: Essays on Kant's Theoretical and Practical Philosophy* (Cambridge: Cambridge University Press, 1996). See also Robert Pippin, "Kant on the Spontaneity of the Mind," in *Idealism as Modernism: Hegelian Variations* (Cambridge: Cambridge University Press, 1997).

21. Strawson, *The Bounds of Sense*, 29.

22. Allison, "Kant's Refutation of Materialism," 94.

23. Ibid., 99. It might be said in reply here that Allison's criticism runs two separate issues together: that of normativity, on the one hand, and that of consciousness, on the other. Among current elaborations of the functionalist positions in cognitive theory, one position attempts to address the issue of normativity by linking the cognitivists' notion of function to the idea of *biological* function, thus importing the type of normativity found in that sphere. See, for example, Ruth G. Millikan, *Language, Thought, and Other Biological Categories* (Cambridge, Mass.: MIT Press, 1984), and Fred Dretske, *Naturalizing the Mind* (Cambridge, Mass.: MIT Press, 1995). In Chapters 5, 6, and 7 I argue that this was a direction essentially anticipated and incorporated within the idealist project by Kant's heirs, Fichte, Schelling, and Hegel.

24. Allison, "Apperception and Analyticity in the B-Deduction," 48.

25. Ibid., 48–49.

26. Kitcher, *Kant's Transcendental Psychology*, 94.

27. As argued, for example, by Gennaro in *Consciousness and Self-Consciousness*.

28. Fred Dretske, *Naturalizing the Mind* (Cambridge, Mass.: MIT Press, 1995), 104–07.

29. Something like this seems to be the view of Andrew Brook who rejects the "logically necessary conditions/psychologism as a false dichotomy" and puts forward a third alternative—"an account that explores *both* the necessary conditions of the mind's operation *and* the actual psychology of these operations and that does the latter precisely by doing the former." Brook, *Kant and the Mind*, 5–6.

30. Kitcher, *Kant's Transcendental Psychology*, 244, note 12.

31. Ibid., 66. With thinkers such as Brentano and Husserl, this notion was fundamentally linked to the idea of a conscious subject *for* whom objects were represented.

32. Ibid., 114 (emphasis added). Kitcher is somewhat tentative here: Kant, she states, "overlooks, or rejects, a *possible* alternative" (ibid., emphasis added) in not treating representation as covariation with appropriate behavior.

33. Kitcher's phrase begs the question of how we are to understand "appropriate." One possibility here would be to employ some biological notion of functional response as suggested by Millikan and Dretske. (See footnote 21.)

34. Ibid., 114.

35. Thomas Nagel, "What Is It Like to Be a Bat?" *Philosophical Review* 83 (1974): 435–50, reprinted in his *Mortal Questions* (Cambridge: Cambridge University Press, 1979).

36. One explicit attempt to do this can be found in Andrew Brook's notion of the apperceiving self as a "self-representing representation" (*Kant and the Mind*, esp. chaps. 3 and 9). I cannot here give Brook's account the treatment it deserves, but will rather signal my concern about the intelligibility of the notion of the mind *as* a representation, self-representing or otherwise.

37. Israel Rosenfield, *The Strange, Familiar, and Forgotten: An Anatomy of Consciousness* (New York: Knopf, 1992), 69.

38. While these cognitive defects were indeed once described in terms of "short-term" versus "long-term" memory, the tendency has been to describe the problem as one of "learning" or "anterograde amnesia" and transfer of information from an intact "working memory" to the short-term memory. See, for example, Raymond D. Adams and Maurice Victor, *Principles of Neurology*, 4th ed. (New York: McGraw-Hill, 1989), 823.

39. This is not quite true. Whereas Rosenfield takes the limbic system to constitute such a level of mental function, LeDoux actually rejects the equation of the emotional brain with the limbic system. Joseph LeDoux, *The Emotional Brain: The Mysterious Underpinnings of Emotional Life* (New York: Simon and Schuster, 1996), 99–103.

40. Rosenfield, *The Strange, Familiar, and Forgotten*, 85.

41. Oliver Sachs, *The Man Who Mistook His Wife for a Hat* (New York: Harper and Row, 1987), 28, quoted in Rosenfield, *The Strange, Familiar, and Forgotten*, 70.

42. Following work by Kihlstrom, LeDoux also believes that "mental representations" of events must be linked to a "mental representation" of the *self* in order for those events to be consciously experienced (*The Emotional Brain*, 279). Again this tends to blur the senses of representation, however.

43. For Fichte's use of the idea of "intellectual intuition" see Chapter 5, especially footnote 12.

Chapter 5

1. Antonio R. Damasio, *Descartes' Error: Emotion, Reason, and the Human Brain* (New York: Putnam, 1994), 150–51.

2. William James, *The Principles of Psychology*, 2 vols. (New York: Dover, 1950), I. 299.

3. Ibid., I. 301–2.

4. Anne Harrington, *Medicine, Mind, and the Double Brain: A Study in Nineteenth-Century Thought* (Princeton: Princeton University Press, 1987), 227–28. Somewhat the same idea appears early in this century with the English neurologist Henry Head's notion of "body schema" (Henry Head, *Studies in Neurology,* vol. 2. [London: Oxford University Press, 1920]). For Head, the body schema was not conscious, but, as Shaun Gallagher has pointed out ("Body Schema and Intentionality," in José Luis Bermúdez, Anthony Marcel, and Naomi Eilan, eds., *The Body and the Self* [Cambridge, Mass.: MIT Press, 1995], 226–28), this notion has often been confused with that of "body image," which *is* conscious.

5. Harrington, *Medicine, Mind, and the Double Brain,* 228–29.

6. Daniel Breazeale, "Check or Checkmate? On the Finitude of the Fichtean Self," in Karl Ameriks and Dieter Sturma, eds., *The Modern Subject: Conceptions of the Self in Classical German Philosophy* (Albany: State University of New York Press, 1995). Reinhard Lauth has stressed this "realist" character of Fichte's thought in *Die Transcendentale Naturlehre Fichtes nach den Principien der Wissenschaftslehre* (Hamburg: Meiner, 1984).

7. Karl L. Reinhold, "The Foundation of Philosophical Knowledge," in George di Giovanni and H. S. Harris, eds., *Between Kant and Hegel* (Albany: State University of New York Press, 1985), 70.

8. Ibid.

9. G. E. Schulze, *Aenesidemus: oder über die Fundamente der von dem Hern Prof. Reinhold in Jena gelieferten Elementar-philosophie, nebst einer Verteidigung gegen die Anmassungen der Vernuftkritik* (Bruxelles: Culture et Civilization, 1969). A translation of part of the essay can be found in Giovanni and H. S. Harris, *Between Kant and Hegel,* 105–35.

10. Neither would Kitcher find Reinhold's account of representation "self-evident." For her, even though a representation might be "distinguished from and related to" an object and even represent its objects "to" some subject, it is clear that this "distinguishing and relating" is not done *by* the subject.

11. J. G. Fichte, "Review of Aenesidemus," in *Fichte: Early Philosophical Writings,* translated and edited by Daniel Breazeale (Ithaca: Cornell University Press, 1988), 59–77.

12. Fichte first used the phrase in "Review of Aenesidemus" of 1794. It does not appear in the *Foundations,* but is used extensively in the "Second Introduction" of 1798. For Fichte, intellectual intuition, qua form of awareness, was an index of the *existence* of that which was presented in it, but it was not a *sensory* awareness (which would be representational).

13. Fichte's "*Wissenschaftslehre*" is usually, but somewhat misleadingly, rendered in English as *The Science of Knowledge.* While the more literal *The Doctrine of Science* is probably preferable, I will use the former translation or, especially in referring to the general project, leave the German untranslated.

14. J. G. Fichte, *The Science of Knowledge,* edited by Peter Heath and John Lachs (Cambridge: Cambridge University Press, 1982), 93.

15. Ibid., 95.

16. John McDowell, *Mind and World* (Cambridge, Mass.: Harvard University Press, 1994), 11.

17. Fichte, *Science of Knowledge,* 38. This more phenomenological approach is also characteristic of the "new method" found in Fichte's lectures of 1796/97, 1797/98, and 1798/99, known from student transcripts and published as J. G. Fichte, *Foundations of Transcendental Philosophy: (Wissenschaftslehre) Nova Methodo (1796/99),* translated and edited by Daniel Breazeale (Ithaca: Cornell University Press, 1992). For an account of Fichte which focuses on the new method, see Günter Zöller, *Fichte's Transcendental Philosophy: The Original Duplicity of Intelligence and Will* (Cambridge: Cambridge University Press, 1998).

18. "Intellectual intuition seems to belong solely to the primordial being, and can never be ascribed to a dependent being, dependent in its existence as well as in its intuition, and

which through that intuition determines its existence solely in relation to given objects." Immanuel Kant, *Critique of Pure Reason*, translated by Norman Kemp Smith (New York: Macmillan, 1929), B 72.

19. Fichte, *Science of Knowledge*, 46.

20. In the nineteenth century, a similar position was held by Helmholtz, a clear precursor to the cognitive scientists of this century.

21. Wilfrid Sellars, "Empiricism and the Philosophy of Mind," in *Science, Perception and Reality* (London: Routledge and Kegan Paul, 1963).

22. McDowell, *Mind and World*, 8.

23. Fichte, *Science of Knowledge*, 246.

24. Breazeale, "Check or Checkmate?," 94.

25. Kant, *Critique of Pure Reason*, A 51, B 75.

26. Fichte, *Science of Knowledge*, 278.

27. F. W. J. von Schelling, *System of Transcendental Idealism* (1800), translated by Peter Heath with an Introduction by Michael Vater (Charlottesville: University of Virginia, 1978), 10.

28. See, for example, Robin May Schott, *Cognition and Eros: A Critique of the Kantian Paradigm* (Boston: Beacon Press, 1988), chap. 8, "Kant's Treatment of Sensibility."

29. Immanuel Kant, *Critique of Judgment*, translated by Werner S. Pluhar (Indianapolis: Hackett, 1987), §7, 55–56.

30. Ibid., 55.

31. Kant, *Critique of Pure Reason*, A 28–29. The A edition is clearer here, but the idea is repeated in the B edition (B 44–45).

32. Kant, *Critique of Judgment*, §3, 48 (emphasis added to phase).

33. Ibid., §5, 51.

34. But for Kant they are not *representations* of the subject: the feeling "is referred solely to the subject (*wird . . . lediglich auf das Subject bezogen*) and is not used for cognition at all, not even for that by which the subject *cognizes* himself (*sich das Subject selbst erkennt*)" Ibid., § 3. Kant's use of *Vorstellung* ("representation") for feelings is an example of those "nonrepresentational" uses of the term referred to by Kitcher. His idea of feeling as a form of nonrepresentational "reference" (*Beziehung*) is analogous to Fichte's idea of feeling as a form of intellectual intuition.

35. Fichte, *Science of Knowledge*, 38 and 39.

36. Ibid., 39–40. For consistency, I have rendered *Vorstellung* throughout as "representation," rather than as "presentation," as is given in the Heath and Lachs translation.

37. As noted, similar ideas giving a primordial role to a type of non-thematic felt-sense of self have been repeated by a variety of psychological theorists from the early nineteenth well into the twentieth century. We might here compare a view put forward in the 1980s by the cognitive scientist Ulric Neisser concerning what he calls one's "ecological self" (Ulric Neisser, "Five Kinds of Self-Knowledge," *Philosophical Psychology* 1 [1988]: 35–59). In a discussion of different types of self-knowledge, Neisser stresses the distinctness of one's immediate and time-dependent sense of self as "a bounded, articulated and controllable body [which] is specified not only by what we can see of it but by what we feel and what we can do" (39). Neisser insists that while we are *conscious* of this ecological self, "awareness of this kind is not what we would ordinarily call 'self-consciousness'." That is, it is not "self-consciousness" understood in the sense of a representational consciousness of the self as an object. "The ecological self *per se* is not an object of thought; very young infants have no internal self-representations to be conscious of or to think about [although they have, he argues, a sense of their ecological selves]. Such representations appear only in extended, private, and conceptual selves. The ecological self, in contrast, is directly perceived" (41). Significantly, within current debates within cognitive science, Neisser is an important critic of the classical representational-computational paradigm, stressing the primordiality of an

embodied, contextualized account of mental function and advocating a Gibsonian (and Jamesian) type of "direct realist" account of perception.

38. Fichte, *Science of Knowledge*, 266.

Chapter 6

1. J. G. Fichte, *The Science of Knowledge*, edited by Peter Heath and John Lachs (Cambridge: Cambridge University Press, 1982), 8.

2. F. W. J. von Schelling, *System of Transcendental Idealism* (Cambridge: Cambridge University Press, 1982), 8.

3. Ibid., 2.

4. See Andrew Bowie, *Schelling and Modern European Philosophy: An Introduction* (London: Routledge, 1993), chap. 4.

5. Schelling, *System of Transcendental Idealism*, 2.

6. Ibid., 15.

7. Ibid., 19 and 27. In this way, the idealist epistemological project to find a ground for knowledge must be kept distinct from the traditional foundationalist project that searches for presuppositionless and certain knowledge of some self-guaranteeing existent able to provide an external ground for all subsequent knowledge. Moreover, the principle sought by the transcendental idealist is a negative one: while seeking "an ultimate of some sort, from which all knowledge begins" the stress here is on the role of this ultimate as a limit, "beyond which there is no knowledge." Such an internalist principle that "even without our being aware of it, absolutely fetters and binds us in knowledge . . . *in the course of our knowing,* never once becomes [a knowable] object, precisely because it is the principle of all knowledge" (ibid., 16). It is this insistent internalism of Schelling's thought in the *System* of 1800 that exempts him from making ontological claims about the primordiality of any self-conscious "I." Because of the nature of the task, our explanations must simply terminate once we reach the principle of self-consciousness: "Why the self should have originally to become aware of itself, is not further explicable, for it is nothing else but self-consciousness" (ibid., 44–45).

8. Ibid., 5–6.

9. Ibid., 57. Again, I have consistently used "representation" for *Vorstellung* (here, in place of the translators' "presentation").

10. Ibid., 6. The theme of the mind's recognition of itself in nature is that found in Kant's *Critique of Judgment.*

11. Ibid., 17. Schelling, it must be admitted, was far from consistent on this, however. The following year, in *Darstellung meines Systems der Philosophie,* he describes himself as employing Spinoza's method of construction. F. W. J. von Schelling, *Schriften,* M. Frank, ed. (Frankfurt am Main: Suhrkamp, 1985), 2, 45.

12. Timothy Lenoir, *The Strategy of Life: Teleology and Mechanics in Nineteenth-Century German Biology* (Dordrecht: Reidel, 1982).

13. In recent philosophy of science such teleological talk of functions is often justified on the post-Darwinian basis that the theory of natural selection has permitted talk of "design without a designer." See, for example, Philip Kitcher, "Function and Design," in P. French, T. Uehling, Jr., and H. Wettstein, eds., *Midwest Studies in Philosophy, 18, Philosophy of Science* (Notre Dame: University of Notre Dame Press, 1993), 380.

14. In cognitive science, the explanatory strategy is sometimes called that of "reverse engineering."

15. Lenoir, *Strategy of Life,* 53. Lenoir demarcates the "teleomechanists" from the supporters of Schellingian nature-philosophy, but this is criticized by Robert Richards in *The Meaning of Evolution: The Morphological Construction and Ideological Reconstruction of*

Darwin's Theory (Chicago: University of Chicago Press, 1992), 48, note. Dov Ospovat gives a clear account of the complexities and philosophical allegiances of these disputes in *The Development of Darwin's Theory: Natural History, Natural Theology, and Natural Selection, 1838–1859* (Cambridge: Cambridge University Press, 1981), chap. 1.

16. See also Karl Figlio, "The Metaphor of Organization in 19th-Century Biomedical Sciences," *History of Science* 13 (1976): 17–53, and L. S. Jacyna, "Immanence and Transcendence: Theories of Life and Organization in Britain, 1790–1835," *Isis* 74 (1983): 311–29.

17. Immanuel Kant, *Critique of Judgment,* translated by Werner S. Pluhar (Indianapolis: Hackett, 1987), § 61, 236. Congruent with this Kant entertained the possibility that in fact biological organization might be produced by the operation of mechanical laws unknowable to us (ibid., § 73, 276–77).

18. Ibid., § 61, 237.

19. Schelling's move here is to reconceive the relation between determinative and reflective judgments. Kant talks as if the world can be conceived from the point of view of a determinative judge, an ideal physicist, who has no reliance on reflective judgments. For Schelling, the determinative judge is a one-sided abstraction of a subject-object and presupposes reflective judgment. Similarly, the physicist's "world" is an abstraction from the much fuller world of that subject-object.

20. Edwin Clarke and L. S. Jacyna, *Nineteenth-Century Origins of Neuroscientific Concepts* (Berkeley: University of California Press, 1987).

21. The term "vegetative nervous system" was that of the Schellingian anatomist J. C. Reil.

22. Clarke and Jacyna, *Nineteenth Century Origins of Neuroscientific Concepts,* 130–33, 129.

23. In modern information theory, the amount of information contained in a sequence of signals is thought of as proportional to the degree to which "uncertainty" or "possibility" is eliminated. That is, the information contained in a flow of signals increases with the degree to which the sequence is *constrained.* Such cybernetic or systems-theoretic ideas did indeed become first applied in the realm of biology to capture the notion of the organism as a homeostatic, or self-regulating, mechanism. It is precisely this idea of a living thing's perpetuation of its own internal states and resistance to equilibration with its environment that is central to Schelling's idea of life: life "must be thought of as engaged in a constant struggle against the course of nature, or in an endeavor to uphold its identity against the latter." Schelling, *System,* 127.

24. Ibid., 122, 124, and 126.

25. For Schelling, such an organized flow of *Vorstellungen* allows the presentation or exhibition (*Darstellung*) of the world within the organized flow itself, an idea that seems to be his version of the Leibnizian idea of the whole of the universe as reflected within the monad. Furthermore, like Leibniz, he wants to have it that different "monadic" intelligences can reflect the same universe to different degrees or in different ways, and he does this by constructing a series of forms of self-consciousness, a "history of self-consciousness in epochs" roughly parallel to those "evolutionary" series of organisms found in contemporary natural history. Thus, for example, Schelling employs Carl Friedrich Kielmayer's classification of the animal world in terms of how the internal structure of different kinds of organisms expressed different configurations among the basic forming forces (*"Bildungskräfte"*) of reproductive power, irritability, and sensibility.

26. Fichte, *Science of Knowledge,* 198–99.

27. Schelling, *System of Transcendental Idealism,* 2.

28. Ibid., 50.

29. On the complexities of the history of the term *evolution,* see Peter Bowler, "The Changing Meaning of 'Evolution'," *Journal of the History of Ideas* 36 (1975): 95–114, and Richards, *The Meaning of Evolution.* The term had in fact first been applied to embryological

development. On the more general use of historical explanations by natural scientists in this period see, for example, Dietrich von Engelhardt, "Historical Consciousness in the German Romantic *Naturforschung*," in Andrew Cunningham and Nicholas Jardine, eds., *Romanticism and the Sciences* (Cambridge: Cambridge University Press, 1990).

30. Schelling, *System of Transcendental Idealism*, 54.

31. "Now since in this act the self opposes to itself the object, the latter will have to appear to it as the negation of all intensity, that is, will have to appear to it as *pure extensity. . . .* But now the intuition whereby inner sense becomes an object to itself is *time . . .* the intuition whereby outer sense becomes an object to itself is *space*," ibid., 103.

32. William James, *The Principles of Psychology*, 2 vols. (New York: Dover, 1950), I. 224.

33. Schelling, *System of Transcendental Idealism*, 102 and 103.

34. On Kielmeyer, see Lenoir, *Strategy of Life*, 37–53. The notion of "irritability" as an intrinsic property of living muscle tissue went back to the work of the Swiss vitalist anatomist Albrecht von Haller in the mid-eighteenth century. On this background see Timothy Lenoir, "The Göttingen School and the Development of Transcendental Naturphilosophie in the Romantic Era," *Studies in the History of Biology* 5 (1981): 111–205.

35. Schelling developed the idea of "potences" or "powers" in his nature philosophy as a way of capturing the type of hierarchical organization he thought typified the organic realm.

36. Schelling, *System of Transcendental Idealism*, 126.

37. Ibid., 11–12.

38. Fichte, *Science of Knowledge*, 38–40.

39. Schelling, *System of Transcendental Idealism*, 43–4.

40. See, for example, J. Laplanche and J.-B. Pontalis, *The Language of Psycho-Analysis* (London: The Hogarth Press, 1973), 333–34 and 427–29.

41. Schelling, *System of Transcendental Idealism*, 65.

42. "The original sensation, in which the self was merely the sensed, is transformed into an intuition, in which the self for the first time becomes for itself that which senses, but for that very reason ceases to be the sensed. For the self that *intuits* itself as sensing, the sensed is the (previously sensing) ideal activity which has crossed the boundary, but is now no longer intuited as an activity of the self." Ibid., 69.

43. For an account of the ubiquity of the principle of polarity in romantic thought and its link to the Schellingian unconscious, see Albert Béguin, *L'Ame romantique et le rêve* (Paris: Corti, 1939).

44. Jacques Lacan, *Ecrits: A Selection*, translated by Alan Sheridan (London: Tavistock, 1977); Ignacio Matte-Blanco, *The Unconscious as Infinite Sets: An Essay on Bi-Logic* (London: Duckworth, 1975), and Eric Rayner, *Unconscious Logic: An Introduction to Matte-Blanco's Bi-Logic and Its Uses* (London: Routledge, 1995).

45. Marcia Cavell, *The Psychoanalytic Mind: From Freud to Philosophy* (Cambridge, Mass.: Harvard University Press, 1993), chap. 8.

Chapter 7

1. Daniel C. Dennett, *Kinds of Minds: Toward on Understanding of Consciousness,* (London: Weidenfeld and Nicholson, 1996), 134–35.

2. Andy Clark, *Being There: Putting Brain, Body, and World Together Again* (Cambridge, Mass.: MIT Press, 1997), 214.

3. Jerry Fodor's hypothesis of a "language of thought" is probably the most explicit and best known of such accounts. See, for example, J. A. Fodor, *The Language of Thought* (New York: Crowell, 1975).

4. Clark, *Being There*, 210.

5. Dennett, *The Intentional Stance.*

6. Hilary Putnam first broached this issue in "The Meaning of Meaning," in *Mind, Language, and Reality: Philosophical Papers,* vol. 2 (Cambridge: Cambridge University Press, 1975). See also Tyler Burge, "Individualism and the Mental," in P. A. French, T. Uehling, and H. Wettstein, eds. *Midwest Studies in Philosophy,* vol. IV (Minneapolis: University of Minnesota Press, 1979).

7. J. G. Fichte, *The Science of Rights,* translated by A. E. Kroeger (Philadelphia: Lippincott, 1869), part 1, "The Deduction of the Conception of Rights."

8. Paul Redding, *Hegel's Hermeneutics* (Ithaca: Cornell University Press, 1996).

9. G. W. F. Hegel, *Hegel's Phenomenology of Spirit,* translated by A. V. Miller (Oxford: Oxford University Press, 1977), § 177 and 178.

10. G. W. F. Hegel, *Elements of the Philosophy of Right,* edited by Allen W. Wood, translated by H. B. Nisbet (Cambridge: Cambridge University Press, 1991), § 158.

11. Ibid., § 164.

12. Daniel Berthold-Bond, *Hegel's Theory of Madness* (Albany: State University of New York Press, 1995).

13. G. W. F. Hegel, *Hegel's Philosophy of Mind: Being Part Three of the Encyclopaedia of Philosophical Sciences,* translated by William Wallace (Oxford: Clarendon Press, 1971), § 408 *zusatz* (translation altered). On this, see Berthold-Bond, *Hegel's Theory of Madness,* chap. 3.

14. J. G. Fichte, *The Science of Knowledge,* edited by Peter Heath and John Lachs (Cambridge: Cambridge University Press, 1982), 278.

15. Hegel, *Philosophy of Mind,* § 401, translation modified.

16. Hegel goes on to talk of the smile as becoming a *gesture* (Ibid., § 401; *zusatz*). Such inner, somatically based feelings humans share, Hegel thinks, with other animals. The peculiarity of their role in human life is bound up with their capacity to function as gestures or signs and for these to become subject to voluntary control. (See Hegel's discussion at § 411 *zusatz*.)

17. Dennett, *The Intentional Stance.*

18. G. W. F. Hegel, *Aesthetics: Lectures on Fine Art,* translated by T. M. Knox (Oxford: Clarendon Press, 1975), 32–33.

19. Ibid., 33.

20. William James, "Place of Affectional Facts in a World of Pure Experience," in *Essays in Radical Empiricism and A Pluralistic Universe* (New York: Longmans, 1947), 80. As for Fichte, for Hegel, the immediacy of sensation or feeling is an occasion for the conceptualized positing of an object deemed responsible for it, a positing that, because of its representational nature, can never fully capture that feeling as such. Moreover, for such an object itself to be determinate its properties must be counterposited with others.

21. Arnold H. Modell, *The Private Self* (Cambridge, Mass.: Harvard University Press, 1993), 22.

22. See especially Jacques Lacan, "The Mirror Stage as Formative of the Function of the I," and "The Function and Field of Speech and Language in Psychoanalysis," both in *Écrits: A Selection,* translated by Alan Sheridan (London: Tavistock, 1977).

23. See, for example, Michel Henry, *The Genealogy of Psychoanalysis,* translated by Douglas Brick (Stanford: Stanford University Press, 1993).

24. Silvan Tomkins, *Shame and Its Sisters: A Silvan Tomkins Reader,* Eve Kosofsky Sedgwick and Adam Frank, eds. (Durham: Duke University Press, 1995), 36. In fact, Tomkins's linking of cybernetics to a more straightforwardly biological approach to the brain reflects the proximity of cybernetics in its earlier phase to the science of physiology. With the notion of "feedback" Norbert Wiener had applied mathematical ideas from the newly developing "information science" to biological ideas of self-regulation that went back into the nineteenth century. In fact, the modern neurological notion of "feedback" has been described as having

its origin in Dewey's development of James's notion of the reflex arc, a development which, according to D. C. Phillips, was influenced by Dewey's earlier *Hegelianism* ("James, Dewey, and the Reflex Arc," *Journal of the History of Ideas* 32 (1971): 566). Be that as it may, the notion seems to have been "in the air," especially among such antimechanistic or "romantic" biological theorists as von Uexküll in the 1920s (Robert McClintock, "Machines and Vitalists: Reflections on the Ideology of Cybernetics," *American Scholar* 35 (1966): 251).

25. Tomkins, *Shame and Its Sisters*, 41: "Just as the experience of redness could not be further described to a color-blind man, so the particular qualities of excitement, joy, fear, sadness, shame, and anger cannot be further described if one is missing the necessary effector and receptor apparatus" (ibid., 41–42).

26. Silvan S. Tomkins, *Exploring Affect: The Selected Writings of Silvan S. Tomkins*, E. Virginia Demos, ed. (Cambridge: Cambridge University Press, 1995), 89–90.

27. Ibid., 285.

28. As Freud's investigations into human sexuality made clear, the experience of a pleasure can become the occasion of a higher-order "pain," and pain, that of a higher-order pleasure.

29. Tomkins, *Exploring Affect*, 81. The psychologist Robin Dunbar has stressed the importance of the practice of mutual grooming in certain primates, an activity that produces endogenous opiates in the groomed partner, and has speculated on the importance of such forms of affect-influencing behavior in the evolution of communication in hominids. Robin Dunbar, *Grooming, Gossip, and the Evolution of Language* (London: Faber and Faber, 1996).

30. Ibid., 79.

31. Ibid., 169.

32. Ibid., 44.

33. Tomkins, *Shame and Its Sisters*, 54.

34. Kant, *Critique of Judgment*, § 59, 226.

35. Ibid., § 59, 228.

36. Schelling, *System of Transcendental Idealism*, 12.

37. "[Art] stands as an exhibition (*Darstellung*) of the infinite on the same level with philosophy; just as philosophy presents the absolute in the *archetype* so also does art present the absolute in a *reflex* or *reflected image*." Friedrich W. J. Schelling, *Philosophy of Art*, edited and translated by Douglas W. Stott (Minneapolis: University of Minnesota Press), 16.

38. Ibid., 3 and 17.

39. Ibid., 30.

40. Ibid., 101 and 99. Schelling's idea of the primacy of "symbolic *Darstellung*" and aesthetio-mythological expression, together with his continuing stress on the "original limitation" and finitude of human existence and thought, found an even more explicit development later in Nietzsche's characterization of "the fundamental human drive to produce images and metaphors" and the primacy that he was to attribute to art and mythology. See Friedrich Nietzsche, "On Truth and Lie in the Extra Moral Sense," in *Philosophy and Truth: Selections from Nietzsche's Notebooks from the Early 1970s*, edited by Daniel Breazeale (Atlantic Highlands, N.J.: Humanities Press, 1979), 88.

41. E. Durkheim and M. Mauss, "De quelque formes primitives de classification," *L'Année sociologique* VI (1901–1902), 1–72, trans. R. Needham as *Primitive Classification* (Chicago: Chicago University Press, 1963), and Robert Hertz, "La Prééminence de la main droite: étude sur la polarité religieuse," *Revue philosophique* 68 (1909), translated by Rodney Needham as "The Pre-eminence of the Right Hand: A Study in Religious Polarity," in Rodney Needham, ed, *Right and Left: Essays on Dual Symbolic Classification* (Chicago: University of Chicago Press, 1973). On the work and influence of Hertz see Robert Parkin, *The Dark Side of Humanity: The Work of Robert Hertz and Its Legacy* (Australia: Harwood Academic Publishers, 1996).

42. Hertz, "The Pre-eminence of the Right Hand," 6.

43. See, for example, Claude Lévi-Strauss, *La Pensée sauvage* (Paris: Librairie Plon, 1962). English translation, *The Savage Mind* (London: Weidenfeld and Nicolson, 1966), 18.

44. See especially Claude Lévi-Strauss, "The Sorcerer and His Magic" and "The Effectiveness of Symbols" in *Structural Anthropology*, translated by Claire Jacobson and Brooke Grundfest Schoepf (Harmondsworth: Penguin Books, 1963).

45. Pierre Bourdieu, *Outline of a Theory of Practice*, translated by Richard Nice (Cambridge: Cambridge University Press, 1977). Rodney Needham especially has pursued this issue. See, for example, his *Counterpoints* (Berkeley: University of California Press, 1987). Bourdieu himself explains such polarities in terms of a "logic of practice."

46. G. E. R. Lloyd, *Polarity and Analogy: Two Types of Argumentation in Early Greek Thought* (Cambridge: Cambridge University Press, 1966). See also, R. A. Prier, *Archaic Logic: Symbol and Structure in Heraclitus, Parmenides, and Empedocles* (The Hague: Mouton, 1976).

47. Bourdieu, *Outline of a Theory of Practice*, 121–22.

48. G. W. F. Hegel, *Hegel's Science of Logic*, translated by A. V. Miller (London: Allen and Unwin, 1969), book 1, chap. 1.

49. G. W. F. Hegel, *Hegels's Logic: Being Part One of the Encyclopedia of The Philosophical Sciences (1830)* translated by Willam Wallce (Oxford: Oxford University Press, 1975), § 90. I have pursue this further in "Hegel's Logic of Being and the Polarities of Presocratic Thought," *The Monist* 74 (1991): 438–56.

50. Fichte, *Science of Knowledge*, 279.

51. Ibid., 282.

52. It is precisely with this type of comprehension that we grasp, for example, that the external symptom of the illness is not the "reality" but a manifestation of it, something that we go beyond in our understanding when we seek its underlying cause.

53. Hegel, *Hegel's Logic*, § 90, *zusatz*.

54. To separate the something's observable qualities from some underlying substance constituting its identity, we would have to be able to ask for the *respect* in which a something differed from its determining qualitative negations. Only then could we say that some particular thing, such as Descartes's piece of wax, had undergone a change in its immediate properties but was, nevertheless, the *same*. But the independent identification of such a property is here impossible. This follows from an application of the basic idea of the contrasting negative now applied to the meta-level of the "respect" in which two things can be said to differ.

55. Thus being and essence, immediacy and mediation, are mutually dependent. Here there seems another parallel with Freud. Jonathan Lear reads Freud as advocating the need for the analysand to not become simply aware of, but to "take responsibility for," the contents of their own unconscious, a dimension of the analysis captured by reversing Freud's slogan to "*Where I am, there it shall be.*" Jonathan Lear, *Love and Its Place in Nature: A Philosophical Interpretation of Freudian Psychoanalysis* (New York: Farrar, Straus and Giroux, 1990), 177–87. Analogously, we might add for Hegel, "Where essence is, there being shall be."

56. See in particular, Robert Pippin, *Hegel's Idealism: The Satisfactions of Self-Consciousness* (Cambridge: Cambridge University Press, 1989), and *Idealism as Modernism*, and Terry Pinkard, *Hegel's Phenomenology: The Sociality of Reason* (Cambridge: Cambridge University Press, 1994).

Chapter 8

1. Edwin Clarke and L. S. Jacyna, *Nineteenth-Century Origins of Neuroscientific Concepts* (Berkeley: University of California Press, 1987), 56.

2. Terrence W. Deacon, *The Symbolic Species: The Co-Evolution of Language and the Brain* (New York: Norton, 1997).

3. In Deacon's account, the onset of such capacities is much earlier than usually posited, a change that has to do with his thesis of the "co-evolution" of language and the brain.

4. Ibid., 55–59. See also Derek Bickerton, *Language and Species* (Chicago: University of Chicago Press, 1990), 12–16.

5. While the uniqueness of this "syntactic" capacity of humans is often stressed (see, for example, Bickerton, *Language and Species*, and William H. Calvin, *How Brains Think: Evolving Intelligence, Then and Now* [New York: Basic Books, 1996], chap. 5), by conceptually linking syntax to the existence of symbolic reference Deacon's approach is more compatible with an "externalist" perspective on mental contents.

6. Deacon, *The Symbolic Species*, 80–92.

7. Ibid., 92.

8. Ibid., 89.

9. The notion of "symbol" is commonly used in other ways, in particular to designate some "motivated" as opposed to "arbitrary" link between token and meaning. I will follow Deacon here in using it in its specifically Peircean way.

10. See Steven Pinker, *The Language Instinct: The New Science of Language and Mind* (New York: William Morrow, 1994), and Bickerton, *Language and Species*. Chomsky himself has always resisted the idea that evolution could explain the appearance of language acquisition in humans. See, for example, Noam Chomsky, *Language and the Problems of Knowledge: The Mamagua Lectures* (Cambridge: MIT Press, 1988). 167.

11. Gerald Edelman, *Bright Air, Brilliant Fire: On the Matter of the Mind* (New York: Basic Books, 1992).

12. Deacon, *The Symbolic Species*, 105. Deacon comments on the similarities of his approach on this count to the work of the Russian L. S. Vygotsky in the 1930s, a psychologist whose work was itself rich in Hegelian insight filtered through the official framework of Marxism (ibid., 451).

13. Ibid., 109.

14. Ibid., 122.

15. Ibid., 321–22.

16. "I think Hegel, especially in the philosophy of objective spirit, uses the notion of right or *rechtlich* much as we would use the notion of norm." Robert B. Pippin, *Idealism as Modernism: Hegelian Variations* (Cambridge: Cambridge University Press, 1997), 388. Given the intrinsically normative nature of signification, it seems the natural place to start from to understand his conception of signification.

17. Immanuel Kant, *The Metaphysical Elements of Justice*, translated by John Ladd (Indianapolis: Bobbs-Merrill, 1965), § 8.

18. G. W. F. Hegel, *Elements of the Philosophy of Right*, edited by Allen W. Wood, translated by H. B. Nisbet (Cambridge: Cambridge University Press, 1991), §§ 57–70.

19. Hegel discusses this "one of the most important notions in philosophy" in *Science of Logic*, translated by A. V. Miller (London: Allen and Unwin, 1969), 106–7.

20. Hegel, *Elements of the Philosophy of Right*, §55.

21. Ibid., § 56.

22. Ibid., § 58.

23. Ibid., § 58 addition, emphasis added.

24. G. W. F. Hegel, *Phenomenology of Spirit*, translated by A. V. Miller (Oxford: Oxford University Press, 1977), § 508.

25. Ibid.

26. John McCumber, *Poetic Interaction: Language, Freedom, Reason* (Chicago: University of Chicago Press, 1989), 53–54.

27. Hegel himself explicitly rejected Lamarckian transmutationism. See, for example, his comments in *Philosophy of Nature (Part Two of the Encyclopaedia of Philosophical Sciences)*,

translated by Michael John Perry, 3 vol. (London: George Allen and Unwin, 1970), vol. 1, § 249. However, he did accept Lamarck's *classification* of animals based as it was on the animal's capacity for feeling, sensibility, and intelligence and his idea of a nontemporal, "transcendental" evolutionary ordering of nature "from the simplest organization to the most perfect, in which nature is the instrument of spirit" (ibid., vol. 3, § 370). See also, Perry's discussion at 366–68.

28. Deacon chooses the biblical quote "and the word became flesh" for the title of key chapter describing the consequences of the acquisition of symbolic reference. For Hegel also, this phrase expressed a profound philosophical truth, albeit, in a metaphorical, mythopoeic way.

29. A similar point is made by the "functionalist" linguist Talmy Givón in *Functionalism and Grammar* (Amsterdam: John Benjamins, 1995.), 439. Givón also adopts Peirce's idea concerning the necessary participation of all three forms of reference in grammatical structure.

30. Deacon, *The Symbolic Species*, 449.

31. Edelman, *Bright Air, Brilliant Fire*, 133.

32. Deacon, *The Symbolic Species*, 400.

33. A somewhat similar suggestion has been made by Robin Dunbar who looks not to sexual partners but to "grooming cliques" as the relevant small-scale relationships crucial for the development of language. Dunbar, *Grooming, Gossip, and the Evolution of Language*.

34. Deacon, *The Symbolic Species*, 313–14.

35. Charles Darwin, *The Expression of the Emotions in Man and Animals* (Chicago: University of Chicago Press, 1965), chaps. 1–3.

36. Darwin here used the sort of idea found in Fechner's "principle of constancy." In England, Carpenter had entertained a somewhat similar theory about the nature of nerve force.

37. Ibid., 57.

38. Ibid., 60.

39. Ibid., 50 (emphasis added).

40. Johann Wolfgang von Goethe, *Theory of Color*, in *Collected Works, Vol. 12: Scientific Studies*, edited and translated by Douglas Miller (Princeton: Princeton University Press, 1995). Goethe's fundamental point was that Newtonian color theory ignored "physiological color," including what we would refer to as the phenomenology of color's "opponent contrasts." That green *looks* opposite to red and blue *looks* opposite to yellow are facts about color that Newtonian theory, with its linear spectrum, does not account for. One of the few theorists of emotion to follow up Darwin's idea of antithetical structuring has been Robert Plutchik who has constructed an "emotion solid" to capture the polar structuring of emotional phenomenology (Robert Plutchik, "Emotion, Evolution, and Adaptive Processes," in Magda B. Arnold, ed., *Feelings and Emotions: The Loyola Symposium* [New York: Academic Press, 1970]).

41. For details of the opponent processing of color see Evan Thompson, *Colour Vision: A Study in Cognitive Science and the Philosophy of Perception* (London: Routledge, 1995), chap. 2.

42. Deacon, *The Symbolic Species*, 419 and 56. Such processes are interpreted as the result of the operation of a particular type of evolutionary logic, "disruptive selection," which favors the extremes of some value range against the intermediate or "compromise" value of a trait (ibid., 56).

43. See Gregory A. Kimball, *Psychology: The Hope of a Science* (Cambridge, Mass.: MIT Press, 1996), chap. 5.

44. As Deacon points out, iconicity cannot be understood simply on the basis of similarity. Rather, like all forms of signification, it is interpretation-relative. Moth wings can be iconic of bark if the degree of resemblance is sufficient to prevent a predatory bird from

distinguishing the moth from the bark (*The Symbolic Species*, 74–75). Nelson Goodman's explorations of the exemplification by a thing of one of its characteristics may provide a clue to the relation of iconicity and opponent structure (Nelson Goodman, *Ways of Worldmaking* [Indianapolis: Hackett Publishing Company, 1978], 63–65.) As Goodman points out, the fact that any single thing has a variety of qualities implies that its ability to exemplify any particular one of them must be a function of context. But the simplest context here would seem to be juxtaposition with a thing having some maximally opposed quality against a background of otherwise shared ones. One grasps a red cup as exemplifying its color when it is juxtaposed to, say, a white one, while in another context, the same cup could be taken as exemplifying its shape. If one understands exemplification as an iconic relation, then opponent process would seem to be crucial for it.

Bibliography

Allison, Henry E. *Idealism and Freedom: Essays on Kant's Theoretical and Practical Philosophy*. Cambridge: Cambridge University Press, 1996.

———. *Kant's Transcendental Idealism: An Interpretation and Defense*. New Haven: Yale University Press, 1983.

Ameriks, Karl, and Dieter Sturma, eds. *The Modern Subject: Conceptions of the Self in Classical German Philosophy*. Albany: State University of New York Press, 1995.

Arens, Katherine. *Structures of Knowing: Psychologies of the Nineteenth Century*. Dordrecht: Kluwer, 1989.

Armstrong, David M. *A Materialist Theory of the Mind*. New York: Humanities Press, 1968.

———. *The Nature of Mind and Other Essays*. Ithaca: Cornell University Press, 1980.

Ashton, Rosemary. *The German Idea: Four English Writers and the Reception of German Thought*. Cambridge: Cambridge University Press, 1980.

———. *G. H. Lewes: A Life*. Oxford: Oxford University Press, 1991.

Barfoot, Michael, ed. *"To Ask the Suffrages of the Patrons": Thomas Laycock and the Edinburgh Chair of Medicine, 1855*. With an Introduction by Michael Barfoot. London: Welcome Institute for the History of Medicine, 1995.

Béguin, Albert. *L'Ame romantique et le rêve*. Paris: Corti, 1939.

Beiser, Frederick. *The Fate of Reason*. Cambridge: Harvard University Press, 1987.

Bermúdez, José Luis, Anthony Marcel, and Naomi Eilan, eds. *The Body and the Self*. Cambridge: MIT Press, 1995.

Berthold-Bond, Daniel. *Hegel's Theory of Madness*. Albany: State University of New York Press, 1995.

Bickerton, Derek. *Language and Species*. Chicago: University of Chicago Press, 1990.

Blackmore J., ed. *Ernst Mach—A Deeper Look*. Dordrecht: Kluwer Academic, 1992.

Block, Ned. "On a Confusion about a Function of Consciousness." *Behavioral and Brain Sciences* 18 (1995): 227–87.

Bourdieu, Pierre. *Outline of a Theory of Practice*. Translated by R. Nice. Cambridge:

Cambridge University Press, 1977.

Bowie, Andrew. *Schelling and Modern European Philosophy: An Introduction.* London: Routledge, 1993.

Breazeale, Daniel. "Check or Checkmate? On the Finitude of the Fichtean Self." In Karl Ameriks and Dieter Sturma, eds., *The Modern Subject: Conceptions of the Self in Classical German Philosophy,* 87–114. Albany: State University of New York Press, 1995.

Broman, Thomas H. *The Transformation of German Academic Medicine, 1750–1820.* Cambridge: Cambridge University Press, 1996.

Brook, Andrew. *Kant and the Mind.* Cambridge: Cambridge University Press, 1994.

Calhoun, Cheshire, and Robert C. Solomon, eds. *What Is an Emotion? Classic Readings in Philosophical Psychology.* New York: Oxford University Press, 1984.

Calvin, William H. *How Brains Think: Evolving Intelligence, Then and Now.* New York: Basic Books, 1996.

Cannon, Walter B. *Bodily Changes in Pain, Hunger, Fear, and Rage.* 2d ed. New York: Appleton, 1929.

Carpenter, William B. *Nature and Man: Essays Scientific and Philosophical.* London: Kegan Paul, Trench & Co., 1888.

Cavell, Marcia. *The Psychoanalytic Mind: From Freud to Philosophy.* Cambridge: Harvard University Press, 1993.

Cekic, Miodrag. "Mach's Phenomenalism as a Link between Physics and Psychology." In *Ernst Mach—A Deeper Look,* edited by J. Blackmore. Dordrecht: Kluwer Academic, 1992.

Chalmers, David J. *The Conscious Mind: In Search of a Fundamental Theory.* Oxford: Oxford University Press, 1996.

Charland, Louis C. "Feeling and Representing: Computational Theory and the Modularity of Affect." *Synthese* 105 (1995): 273–301.

Clark, Andy. *Being There: Putting Brain, Body, and World Together Again.* Cambridge: MIT Press, 1997.

Clarke, Edwin, and L. S. Jacyna. *Nineteenth-Century Origins of Neuroscientific Concepts.* Berkeley: University of California Press, 1987.

Cummins, Robert. "Functional Analysis." *Journal of Philosophy* 72 (1975): 741–65.

Cunningham, Andrew, and Nicholas Jardine. *Romanticism and the Sciences.* Cambridge: Cambridge University Press, 1990.

Damasio, Antonio R. *Descartes' Error: Emotion, Reason, and the Human Brain.* New York: Putnam's, 1994.

Danziger, Kurt. *Naming the Mind: How Psychology Found Its Language.* London: Sage Publications, 1997.

Darwin, Charles. *The Expression of the Emotions in Man and Animals.* Chicago: University of Chicago Press, 1965.

Davidson, Richard J. "Complexities in the Search for Emotion-Specific Physiology." In Paul Ekman and Richard J. Davidson, eds., *The Nature of Emotion: Fundamental Questions.* New York: Oxford University Press, 1994.

Deacon, Terrence W. *The Symbolic Species: The Co-Evolution of Language and the Brain.* New York: Norton, 1997.

Dennett, Daniel C. *The Intentional Stance.* Cambridge: MIT Press, 1987.

———. *Kinds of Minds: Towards an Understanding of Consciousness.* London: Weidenfeld and Nicholson, 1996.

Desmond, Adrian, and James Moore. *Darwin.* London: Michael Joseph, 1991.

Dewey, John. "The Reflex Arc Concept in Psychology." *Psychological Review* 3 (1896): 357–70.

Dewhurst, Kenneth. *Hughlings Jackson on Psychiatry.* Oxford: Sandford, 1982.
Doerner, Klaus. *Madmen and the Bourgeoisie: A Social History of Insanity and Psychiatry.* Oxford: Blackwell, 1981.
Dretske, Fred. "The Intentionality of Cognitive States." In David M. Rosenthal, ed., *The Nature of Mind.* New York: Oxford University Press, 1991.
——. *Naturalizing the Mind.* Cambridge: MIT Press, 1995.
Dunbar, Robin. *Grooming, Gossip, and the Evolution of Language.* London: Faber and Faber, 1996.
Durkheim, Emile, and Marcel Mauss. *"De quelque formes primitives de classification."* *L'Année sociologique* VI (1901–1902): 1–72. Translated by R. Needham as *Primitive Classification.* Chicago: Chicago University Press, 1963.
Edelman, Gerald. *Bright Air, Brilliant Fire: On the Matter of the Mind.* New York: Basic Books, 1992.
Ekman, Paul, and Richard J. Davidson, eds. *The Nature of Emotion: Fundamental Questions.* New York: Oxford University Press, 1994.
Ellenberger, Henri F. *The Discovery of the Unconscious: The History and Evolution of Dynamic Psychiatry.* New York: Basic Books, 1970.
——. "Fechner and Freud." In Mark S. Micale, ed., *Beyond the Unconscious.* Princeton: Princeton University Press, 1993.
Fechner, Gustav T. *Elemente der Psychophysik.* Leipzig: Brietkopf und Härtel, 1860. Translated by Helmut E. Adler as *Elements of Psychophysics.* Edited by David H. Howes and Edwin G. Boring. New York: Holt, Rinehart and Winston, 1966.
Feigl, Herbert. "The 'Mental' and the 'Physical'." *Minnesota Studies in the Philosophy of Science* 2 (1958): 370–497.
Fichte, J. G. *Early Philosophical Writings.* Translated and edited by Daniel Breazeale. Ithaca: Cornell University Press, 1988.
——. *Foundations of Transcendental Philosophy (Wissenschaftslehre) Nova Methodo.* Translated and edited by Daniel Breazeale. Ithaca: Cornell University Press, 1992.
——. *The Science of Knowledge.* Edited by Peter Heath and John Lachs. Cambridge: Cambridge University Press, 1982.
——. *The Science of Rights.* Translated by A. E. Kroeger. Philadelphia: Lippincott, 1869.
Figlio, Karl. "The Metaphor of Organization in 19th-Century Biomedical Sciences." *History of Science* 13 (1976): 17–53.
Fodor, J. A. "Fodor's Guide to Mental Representation." In Stephen P. Stich and Ted A. Warfield, eds., *Mental Representation: A Reader.* Oxford: Blackwell, 1990.
——. *The Language of Thought.* New York: Crowell, 1975.
——. *The Modularity of Mind.* Cambridge: MIT Press, 1983.
Forrester, John. *Language and the Origins of Psychoanalysis.* London: Macmillan, 1980.
Freud, Sigmund. *On Aphasia.* New York: International University Press, 1953.
——. *The Standard Edition of the Complete Psychological Works of Sigmund Freud.* Translated and edited by James Strachey. London: Hogarth Press, 1981.
Gallagher, Shaun. "Body Schema and Intentionality." In José Luis Bermúdez, Anthony Marcel, and Naomi Eilan, eds., *The Body and the Self.* Cambridge: MIT Press, 1995.
Galton, Francis. *Inquiries into Human Faculty.* London: Dent, 1907.
Gardner, Howard. *The Mind's New Science: A History of the Cognitive Revolution.* New York: Basic Books, 1985.
Gennaro, Rocco J. *Consciousness and Self-Consciousness: A Defense of the Higher-Order Thought Theory of Consciousness.* Amsterdam: John Benjamins, 1995.

Givón, Talmy. *Functionalism and Grammar*. Amsterdam: John Benjamins, 1995.

Goethe, Johann Wolfgang von. *Theory of Color*. In *Collected Works, Vol. 12: Scientific Studies*. Edited and translated by Douglas Miller. Princeton: Princeton University Press, 1995.

Goldstein, R. G. "The Higher and Lower in Mental Life: An Essay on J. Hughlings Jackson and Freud." *Journal of the American Psychoanalytic Association* 43 (1995): 495–515.

Gordon, Robert. "The Aboutness of Emotions." *American Philosophical Quarterly* II (1974): 27–36.

Griffiths, Paul E. *What Emotions Really Are: The Problem of Psychological Categories*. Chicago: University of Chicago Press, 1997.

Harrington, Anne. *Medicine, Mind, and the Double Brain: A Study in Nineteenth-Century Thought*. Princeton: Princeton University Press, 1987.

Hartmann, Eduard von. *The Philosophy of the Unconscious*. Translated by W. C. Coupland. London: English and Foreign Philosophical Library, 1884.

Hatfield, Gary C. *The Natural and the Normative: Theories of Spatial Perception from Kant to Helmholtz*. Cambridge: MIT Press, 1990.

Head, Henry. *Studies in Neurology*, vol. 2. London: Oxford University Press, 1920.

Hegel, G. W. F. *Aesthetics: Lectures on Fine Art*. Translated by T. M. Knox. Oxford: Clarendon Press, 1975.

———. *Elements of the Philosophy of Right*. Edited by Allen W. Wood, translated by H. B. Nisbet. Cambridge: Cambridge University Press, 1991.

———. *Hegel's Logic: Being Part One of the Encyclopaedia of The Philosophical Sciences (1830)*. Translated by William Wallace. Oxford: Oxford University Press, 1975.

———. *Phenomenology of Spirit*. Translated by A. V. Miller. Oxford: Oxford University Press, 1977.

———. *Philosophy of Mind: Being Part Three of the Encyclopaedia of Philosophical Sciences*. Translated by William Wallace. Oxford: Clarendon Press, 1971.

———. *Philosophy of Nature (Part Two of the Encyclopaedia of Philosophical Sciences)*. Translated by Michael John Perry, 3 vols. London: George Allen and Unwin, 1970.

———. *Science of Logic*. Translated by A. V. Miller. London: Allen and Unwin, 1969.

Henry, Michel. *The Genealogy of Psychoanalysis*. Translated by Douglas Brick. Stanford: Stanford University Press, 1993.

Hertz, Robert. "*La Prééminence de la main droite: étude sur la polarité religieuse,*" *Rev philosophique* 68 (1909). Translated by R. Needham as "The Pre-eminence of the Right Hand: A Study in Religious Polarity." In R. Needham, ed., *Right and Left: Essays on Dual Symbolic Classification*. Chicago: University of Chicago Press, 1973.

Humphrey, Nicholas. "Blocking Out the Distinction between Sensation and Perception: Superblindsight and the Case of Helen." *Behavioral and Brain Sciences* 18 (1995): 257–58.

———. *A History of the Mind*. New York: Simon and Schuster, 1992.

Jackson, John Hughlings. *Selected Writings of John Hughlings Jackson*, 2 vols. London: Hodder and Soughton, 1931.

Jacyna, L. S. "Immanence and Transcendence: Theories of Life and Organization in Britain, 1790–1835." *Isis* 74 (1983): 311–29.

Jaeger, Sigfried. "Fechners Psychophysik im Kontext seiner Weltanschauung." In *G. T. Fechner and Psychology*. Edited by Josef Brozek and Horst Gundlach. Passau: Passavia Universitätsverlag, 1988.

James, William. *Essays in Psychology*. Cambridge: Harvard University Press, 1983.

———. *Essays in Radical Empiricism and A Pluralistic Universe*. New York: Longmans, 1947.

——. *Pragmatism and The Meaning of Truth*. Introduction by A. J. Ayer. Cambridge: Harvard University Press, 1975.

——. *The Principles of Psychology*, 2 vols. New York: Dover, 1950.

Jardine, Nicholas. "*Naturphilosophie* and the Kingdoms of Nature." In N. Jardine, J. A. Secord, and E. C. Spary, eds., *Cultures of Natural History*. Cambridge: Cambridge University Press, 1996.

Jung, Carl Gustav. "Die Psychopathologische Bedeutung des Assoziationsexperimentes." *Archiv für Kriminal-Anthropologie und Kriminalistik* 22 (1906): 145–62.

Kant, Immanuel. *Anthropology from a Pragmatic Point of View*. Translated by Victor Lyle Dowdell, with an introduction by Frederick P. Van De Pitte. Carbondale: Southern Illinois University Press, 1978.

——. *Critique of Judgment*. Translated by Werner S. Pluhar. Indianapolis: Hackett, 1987.

——. *Critique of Pure Reason*. Translated by N. Kemp Smith. New York: Macmillan, 1929.

——. *Metaphysical Foundations of Natural Science*. Translated by James W. Ellington. In Immanuel Kant, *Philosophy of Material Nature*. Indianapolis: Hackett, 1985.

Kenny, Anthony. *Action, Emotion, and Will*. London: Routledge and Kegan Paul, 1963.

Kerr, John. *A Most Dangerous Method: The Story of Jung, Freud, and Sabina Spielrein*. New York: Knopf, 1993.

Kimball, Gregory A. *Psychology: The Hope of a Science*. Cambridge: MIT Press, 1996.

Kitcher, Patricia. "Kant's Real Self." In *Self and Nature in Kant's Philosophy*. Edited by Allen W. Wood. Ithaca: Cornell University Press, 1984.

——. *Kant's Transcendental Psychology*. Oxford: Oxford University Press, 1990.

Kitcher, Philip. "Function and Design." In P. French, T. Uehling, Jr., and H. Wettstein, eds., *Midwest Studies in Philosophy 18, Philosophy of Science*. Notre Dame: University of Notre Dame Press, 1993.

Klemm, David E., and Günter Zöller, eds. *Figuring the Self: Subject, Absolute, and Others in Classical German Philosophy*. Albany: State University of New York Press, 1997.

Lacan, Jacques. *Ecrits: A Selection*. Translated by Alan Sheridan. London: Tavistock, 1977.

Laplanche, J., and J.-B. Pontalis. *The Language of Psycho-Analysis*. London: The Hogarth Press, 1973.

Lazarus, R. S. "On the Primacy of Cognition." *American Psychologist* 39 (1984): 124–129.

Lazarus, Richard S., and Bernice N. Lazarus. *Passion and Reason: Making Sense of our Emotions*. New York: Oxford University Press, 1994.

Lear, Jonathan. *Love and Its Place in Nature: A Philosophical Interpretation of Freudian Psychoanalysis*. New York: Farrar, Straus and Giroux, 1990.

Leary, David E. "The Philosophical Development of the Concept of Psychology in Germany, 1780–1850." *Journal of the History of the Behavioural Sciences* 14 (1978): 113–21.

LeDoux, Joseph. *The Emotional Brain: The Mysterious Underpinnings of Emotional Life*. New York: Simon and Schuster, 1996.

Lennig, Petra. "Gustav Fechner und die Naturphilosophie." In Josef Brozek and Horst Gundlach, eds., *G. T. Fechner and Psychology*. Passau: Passavia Universitätsverlag, 1988.

Lenoir, Timothy. "The Göttingen School and the Development of Transcendental

Naturphilosophie in the Romantic Era." *Studies in the History of Biology* 5 (1981): 111–205.
——. *The Strategy of Life: Teleology and Mechanism in Nineteenth-Century German Biology.* Dordrecht: Reidel, 1982.
Lévi-Strauss, Claude. *La Pensée Sauvage.* Paris: Librairie Plon, 1962. Translated as *The Savage Mind.* London: Weidenfeld, and Nicolson, 1966.
——. *Structural Anthropology.* Translated by Claire Jacobson and Brooke Grundfest Schoepf. Harmondsworth: Penguin Books, 1963.
Lloyd, G. E. R. *Polarity and Analogy: Two Types of Argumentation in Early Greek Thought.* Cambridge: Cambridge University Press, 1966.
Mach, Ernst. "Lectures on Psychophysics—Conclusion." In J. Blackmore, ed., *Ernst Mach—A Deeper Look.* Dordrecht: Kluwer, 1992.
MacLean, Paul. "Psychosomatic Disease and the 'Visceral Brain': Recent Developments Bearing on the Papez Theory of Emotion." *Psychosomatic Medicine* 11 (1949): 338–53.
Magoun, Horace. "Evolutionary Concepts of Brain Function Following Darwin and Spencer." In Sol Tax, ed., *Evolution after Darwin, Vol. II, The Evolution of Man.* Chicago: University of Chicago Press, 1960.
Marshall, M. E. "G. T. Fechner: Premises toward a General Theory of Organisms (1823)." *Journal of the History of the Behavioural Sciences* 10 (1974): 438–47.
Matte-Blanco, Ignacio. *The Unconscious as Infinite Sets: An Essay on Bi-Logic.* London: Duckworth, 1975.
McClintock, Robert. "Machines and Vitalists: Reflections on the Ideology of Cybernetics." *American Scholar* 35 (1966): 249–57.
McCumber, John. *Poetic Interaction: Language, Freedom, Reason.* Chicago: University of Chicago Press, 1989.
McDowell, John. *Mind and World.* Cambridge: Harvard University Press, 1994.
Meares, Russell. *The Metaphor of Play: Disruption and Restoration in the Borderline Experience.* Northvale, N.J.: Jason Aronson, 1993.
Meerbote, Ralf. "Apperception and Objectivity." *Southern Journal of Philosophy*, vol. 25 supp (1987): 115–30.
——. "Kant's Functionalism." In J-C Smith, ed., *Historical Foundations of Cognitive Science.* Dordrecht: Kluwer, 1990.
Metzinger, Thomas, ed. *Conscious Experience.* Paderborn: Schöningh, 1995.
Meynert, Theodor. *Psychiatry: A Clinical Treatise on Diseases of the Fore-Brain Based upon a Study of Its Structure, Functions, and Nutrition.* Translated by B. Sachs. New York: Putnam's, 1885.
Millikan, Ruth G. *Language, Thought, and Other Biological Categories.* Cambridge: MIT Press, 1984.
Modell, Arnold H. *The Private Self.* Cambridge: Harvard University Press, 1993.
Morgan, S. R. "Schelling and the Origin of His Nature Philosophy." In Andrew Cunningham and Nicholas Jardine, eds., *Romanticism and the Sciences.* Cambridge: Cambridge University Press, 1990.
Myers, Gerald E. *William James, His Life and Thought.* New Haven: Yale University Press, 1986.
Nagel, Thomas. "What Is It Like to Be a Bat?" *Philosophical Review* 83 (1974): 435–50. Reprinted in *Mortal Questions.* Cambridge: Cambridge University Press, 1979.
Natsoulas, Thomas. "Conscious Perception and the Paradox of 'Blind-sight'." In G. Underwood, ed., *Aspects of Consciousness, vol 3. Awareness and Self-Consciousness.* London: Academic Press, 1982.
Needham, Rodney. *Counterpoints.* Berkeley: University of California Press, 1987.

——, ed. *Right and Left: Essays on Dual Symbolic Classification*. Chicago: University of Chicago Press, 1973.

Neisser, Ulric. "Five Kinds of Self-Knowledge." *Philosophical Psychology* 1 (1988): 35–59.

Nelkin, Norton. "Propositional Attitudes and Consciousness." *Philosophy and Phenomenological Research* 49 (1989): 413–30.

——. "Unconscious Sensations." *Philosophical Psychology* 2 (1989): 129–141.

Neu, Jerome, ed. *The Cambridge Companion to Freud*. Cambridge: Cambridge University Press, 1991.

Neuhouser, Frederick. *Fichte's Theory of Subjectivity*. Cambridge: Cambridge University Press, 1990.

Nietzsche, Friedrich. "On Truth and Lie in the Extra Moral Sense." In Daniel Breazeale, ed., *Philosophy and Truth: Nietzsche's Notebooks from the Early 1870s*. Atlantic Highlands, N.J.: Humanities Press, 1979.

Noll, Richard. *The Jung Cult: Origins of a Charismatic Movement*. Princeton: Princeton University Press, 1994.

Ospovat, Dov. *The Development of Darwin's Theory: Natural History, Natural Theology, and Natural Selection, 1838–1859*. Cambridge: Cambridge University Press, 1981.

Papez, J. W. "A Proposed Mechanism of Emotion." *Archives of Neurology and Psychiatry* 79 (1937): 217–24.

Peters, Uwe Henrik. *Studies in German Romantic Psychiatry: Justinus Kerner as a Psychiatric Practitioner, E. T. A. Hoffmann as a Psychiatric Theorist*. London: Institute of German Studies, 1990.

Phillips, D. C. "James, Dewey, and the Reflex Arc." *Journal of the History of Ideas* 32 (1971): 555–68.

Pinkard, Terry. *Hegel's Phenomenology: The Sociality of Reason*. Cambridge: Cambridge University Press, 1994.

Pinker, Steven. *The Language Instinct: The New Science of Language and Mind*. New York: Morrow, 1994.

Pippin, Robert B. *Hegel's Idealism: The Satisfactions of Self-Consciousness*. Cambridge: Cambridge University Press, 1989.

——. *Idealism as Modernism: Hegelian Variations*. Cambridge: Cambridge University Press, 1997.

Place, U. T. "Is Consciousness a Brain Process?" *British Journal of Psychology* 47 (1956): 44–50.

Plutchik, Robert. "Emotion, Evolution, and Adaptive Processes." In Magda B. Arnold, ed., *Feelings and Emotions: The Loyola Symposium*. New York: Academic Press, 1970.

Prier, R. A. *Archaic Logic: Symbol and Structure in Heraclitus, Parmenides, and Empedocles*. The Hague: Mouton, 1976.

Putnam, Hilary. *Mind, Language and Reality*. Philosophical Papers, vol. 2. Cambridge: Cambridge University Press, 1975.

——. *Realism and Reason*. Philosophical Papers, vol. 3. Cambridge: Cambridge University Press, 1983.

——. *Representation and Reality*. Cambridge: MIT Press, 1989.

Redding, Paul. *Hegel's Hermeneutics*. Ithaca: Cornell University Press, 1996.

——. "Hegel's Logic of Being and the Polarities of Presocratic Thought." *The Monist* 74 (1991): 438–56.

Reed, Edward S. *From Soul to Mind: The Emergence of Psychology from Erasmus Darwin to William James*. New Haven: Yale University Press, 1997.

——. *James J. Gibson and the Psychology of Perception*. New Haven: Yale University Press, 1988.

Rehbock, Philip F. *Philosophical Naturalists: Themes in Early-Nineteenth-Century British Biology*. Madison: University of Wisconsin Press, 1983.

Richards, Robert J. *Darwin and the Emergence of Evolutionary Theories of Mind and Behavior*. Chicago: University of Chicago Press, 1987.

——. *The Meaning of Evolution: The Morphological Construction and Ideological Reconstruction of Darwin's Theory*. Chicago: University of Chicago Press, 1992.

Ritvo, Lucille B. *Darwin's Influence on Freud: A Tale of Two Sciences*. New Haven: Yale University Press, 1990.

Rizzuto, A.-M. "Freud's Speech Apparatus and Spontaneous Speech." *International Journal of Psychoanalysis* 74 (1993): 113–27.

——. "The Origins of Freud's Concept of Object Representation (*Objektvorstellung*) in His Monograph 'On Aphasia,' Its Theoretical and Technical Importance." *International Journal of Psychoanalysis* 71 (1990): 241–48.

Rock, Irvin. "A Look back at William James's Theory of Perception." In Michael G. Johnson and Tracy B. Henley, eds., *Reflections on the Principles of Psychology: William James after a Century*. Hillsdale, N.J.: Erlbaum Associates, 1990.

Rorty, Amélie. "Explaining Emotions." In *Explaining Emotions*. Berkeley: University of California Press, 1980.

Rosenfield, Israel. *The Strange, Familiar, and Forgotten: An Anatomy of Consciousness*. New York: Knopf, 1992.

Rosenthal, David M. "Higher-Order Thoughts and the Appendage Theory of Consciousness." *Philosophical Psychology* 6 (1993): 155–67.

——. *The Nature of Mind*. New York: Oxford University Press, 1991.

——. "Two Concepts of Consciousness." *Philosophical Studies* 49 (1986): 329–59.

Ryle, Gilbert. *The Concept of Mind*. London: Hutchinson, 1949.

Sachs, Oliver. *The Man Who Mistook His Wife for a Hat*. New York: Harper and Row, 1987.

Schelling, F. W. J. von. *On the History of Modern Philosophy*. Translated and with an introduction by Andrew Bowie. Cambridge: Cambridge University Press, 1994.

——. *Philosophy of Art*. Edited and translated by Douglas W. Stott. Minneapolis: University of Minnesota Press, 1989.

——. *Schriften*. Edited by Manfred Frank. Frankfurt am Main: Suhrkamp, 1985.

——. *System of Transcendental Idealism (1800)*. Translated by Peter Heath with an Introduction by Michael Vater. Charlottesville: University Press of Virginia, 1978.

Schott, Robin May. *Cognition and Eros: A Critique of the Kantian Paradigm*. Boston: Beacon Press, 1988.

Schubert, Gotthilf Heinrich. *Die Symbolik des Traumes*. Faksimiledruck nach der Ausgabe von 1814, mit einem Nachwort von Gerhard Sauder. Heidelberg: Verlag Lambert Schneider, 1968.

Sellars, Wilfrid. "Empiricism and the Philosophy of Mind." In *Science, Perception, and Reality*. London: Routledge and Kegan Paul, 1968.

——. "Kant's Views on Sensibility and Understanding." *The Monist*, 51 (1967): 463–91.

——. "Metaphysics and the Concept of a Person." In Karel Lambert, ed. *The Logical Way of Doing Things*. Yale University Press, 1969.

——. "This I or He or It (This Thing) Which Thinks." Presidential Address to the Eastern Division of the APA, 1970.

Slack, C. W. "Feedback Theory and the Reflex Arc Concept." *Psychological Review* 62 (1955): 263–67.

Sloan, Phillip R. "Introduction" to *Richard Owen. The Hunterian Lectures in Comparative Anatomy, May–June, 1937. With an Introductory Essay and Commentary*. Edited by Phillip R. Sloan. Chicago: University of Chicago Press, 1992.

Sloman, Aaron. "Motives, Mechanisms, and Emotions." In Margaret A. Boden, ed., *The Philosophy of Artificial Intelligence*. Oxford: Oxford University Press, 1990.

Smart, J. J. C. "Sensations and Brain Processes." *Philosophical Review* 68 (1959): 141–56.

Solomon, Robert. *The Passions*. Notre Dame: University of Notre Dame Press, 1983.

Spencer, Herbert. *Principles of Psychology*, 2d ed. London: Williams and Norgate, 1870.

Stocker, Michael. "Emotional Thoughts," *American Philosophical Quarterly* 24 (1987): 59–69.

Straumann, Heinrich. *Justinus Kerner und der Okkultismus in der Deutschen Romantik*. Leipzig: Verlage der Münster-Presse, Horgen-Zürich, 1928.

Strawson, P. F. *The Bounds of Sense: An Essay on Kant's "Critique of Pure Reason."* London: Methuen, 1966.

——. *Individuals*. London: Methuen, 1959.

Sulloway, Frank J. *Freud, Biologist of the Mind: Beyond the Psychoanalytic Legend*. New York: Basic Books, 1979.

Tomkins, Silvan S. *Exploring Affect: The Selected Writings of Silvan S. Tomkins*. Edited by E. Virginia Demos. Cambridge: Cambridge University Press, 1995.

——. *Shame and Its Sisters: A Silvan Tomkins Reader*. Edited by Eve Kosofsky Sedgwick and Adam Frank. Durham: Duke University Press, 1995.

Tye, Michael. *Ten Problems of Consciousness: A Representational Theory of the Phenomenal Mind*. Cambridge: MIT Press, 1995.

Whyte, Lancelot Law. *The Unconscious before Freud*. London: Associated Book Publishers, 1967.

Wollheim, Richard, and James Hopkins, eds. *Philosophical Essays on Freud*. Cambridge: Cambridge University Press, 1982.

Zajonc, R. B. "Emotions, Cognition, Behaviour." *American Psychologist* 93 (1980): 151–55.

——. "Feeling and Thinking: Preferences Need No Inferences." *American Psychologist* 35 (1980): 151–75.

——. "On the Primacy of Affect." *American Psychologist* 39 (1984): 117–23.

Ziolkowski, Theodore. *German Romanticism and Its Institutions*. New Haven: Yale University Press, 1990.

Zöller Günter. *Fichte's Transconducted Philosophy: The Original Duplicity of Intelligence and Will*. Cambridge: Cambridge University Press, 1998.

Index